Sandstone Depositional Models for Exploration for Fossil Fuels

Sandstone Depositional Models for Exploration for Fossil Fuels

Third Edition

George deVries Klein

PROFESSOR OF GEOLOGY
UNIVERSITY OF ILLINOIS AT URBANA-CHAMPAIGN

INTERNATIONAL HUMAN
RESOURCES DEVELOPMENT
CORPORATION
Boston

Cover illustration: Horizontal seismic section at depth of 196 msec showing meandering channel comprising part of Birdfoot Delta, the upper delta plain of which appears in uppermost center, Gulf of Thailand. (Color photograph courtesy of A. R. Brown; from Brown, A. R., Graebner, R. J., and Dahm, C. G., 1982, Use of horizontal seismic sections to identify subtle traps: in Halbouty, M. T., ed., 1982, The deliberate search for the subtle trap: Am. Assoc. Petroleum Geol. Mem. 32, 47–56; republished with permission of the American Association of Petroleum Geologists.) See also Figure 5.33 in text.

Cover and interior design by Outside Designs

Library of Congress Cataloging in Publication Data
Klein, George deVries, 1933–
   Sandstone depositional models for exploration for fossil fuels.

   Bibliography: p. 183
   Includes index.
   1. Prospecting. 2. Sandstone. 3. Petroleum—Geology. 4. Gas, Natural—Geology. 5. Uranium ores. I. Title.
TN271.P4K58   1985   553.2′8   84–29735
ISBN 0–934634–82–3
[ISBN 90–277–2064–9 D. Reidel]

Printed in the United States of America

Geological Sciences Series

Contents

Acknowledgments

This monograph is an outgrowth of several short courses I have presented dealing with sandstone models and exploration. The original stimulus for preparing the first edition of this monograph came from Dr. Ram S. Saxena of New Orleans in 1974. I would like to take this opportunity to thank Dr. Saxena once again for his encouragement and for later providing me with material that was difficult to obtain.

Since 1980, when the second edition was published, many new developments have required reappraisal of the earlier edition. Although most of the basic principles of depositional systems described in the second edition are still valid, new sedimentological concepts have emerged in the intervening period, and much newly released information has been published dealing with the application of sandstone models to exploration for oil, coal, uranium, and mineral deposits. Because these were incorporated into my short course lectures on a continuing basis, the lecture content had outstripped the materials in the second edition. Therefore, this revision was undertaken to align lecture content much more closely with this monograph.

Permission to reprint illustrations was granted by the University of Chicago Press, the Burgess Publishing Company, the International Association of Sedimentologists, the Geological Society of America, the Netherlands Geological and Mining Society, the Geological Society of Italy, the Canadian Society of Petroleum Geologists, the Society of Exploration Geophysicists, the European Society of Exploration Geophysicists, the Society of Economic Paleontologists and Mineralogists, the Society of Economic Geologists, Dowden Hutchinson and Ross, Inc., John Wiley and Sons, Elsevier Publishing Co., and the American Association of Petroleum Geologists. I also wish to thank the many authors who granted me permission to reprint their illustrations and provided original copy; appropriate acknowledgment to them is given in the figure captions.

Finally there are others to whom I wish to express my appreciation for their assistance. First, I wish to thank the many past participants in my short courses for their stimulating and probing questions that helped focus many of the concepts presented in this monograph. Second, I wish to thank Michael R. Hays, Publisher, and his staff at IHRDC for facilitating in so many ways the revision of this monograph. Third, I thank Joanne Klussendorf (Mikulic) for drafting many of the illustrations, and Mrs. Terri Monnett for typing the manuscript. Finally, I thank Michael A. Arthur for his careful review.

George deVries Klein
Urbana, Illinois
November 15, 1984

Sandstone Depositional Models for Exploration for Fossil Fuels

Chapter 1

Introduction

The purpose of this monograph is to provide a brief summary of the depositional processes, Holocene sediments, ancient counterparts, and examples of oil- and gas-bearing stratigraphic traps in fifteen clastic depositional environments. This summary is intended to complement lecture and reading courses dealing with sedimentology, depositional systems, sedimentary facies, sedimentary environments, sandstone diagenesis, and sedimentary modeling as predictive tools for exploration. The student is cautioned, however, that this monograph is merely an introduction and summary overview of the subject and no claim is made for topical or bibliographic completeness. More complete treatments are found in standard textbooks.

The field of sedimentology has changed and advanced over the past thirty years, partly as a result of the stimulus of the American oil industry, which needed to make predictions about the occurrence of harder-to-find stratigraphic traps. In addition, with the emergence of plate tectonic theory and the supportive data from the Deep Sea Drilling Project (DSDP), the field of sedimentology moved from essentially a descriptive science to a mature, predictive science. The 1960s, 1970s, and 1980s in particular witnessed an explosion of new insights and understanding of how sediments are deposited, and how sedimentary rocks are formed and altered.

The approach that emerged is based on the recognition that sedimentary rocks owe their origin to physical and chemical processes that can be cast into process–response models (Table 1.1). It is now known that sediments and sedimentary rocks are the product of interactions of sed-

TABLE 1.1. *Flow sheet for predictive process-response models in sedimentology*

Process	+	Material	→	Response	→	Prediction
Open-channel flow		Texture		Sedimentary structures		Environment of
				Vertical sequence		deposition
Wave systems				Geometry		Diagenetic changes
				Porosity		
Tidal currents		Mineralogy		Permeability		Updip direction
Jet flow				SP log response		Downdip direction
Debris flow				Resistivity log response		Depositional strike
Slump				Gamma-ray log response		Oil
Turbidity current				Sonic velocity		Natural gas
———				Seismic reflector		Coal
Pressure						Uranium
Temperature						Iron ore
						Placer gold
						Copper
						Lead
						Zinc

imentary materials consisting of particles of different sizes and minerals, with various transport and depositional processes (rivers, tidal currents, waves, for instance) and chemical processes that generate a characteristic response that is preserved both in Holocene sediments and the rock record. This response consists of sedimentary structures, the vertical sequence (of grain size changes, lithologies and sedimentary structures), the geometry of sedimentary bodies, the properties of primary porosity and permeability, and secondary porosity. Shapes of electric, gamma-ray and sonic velocity logs, and the nature of seismic reflectors are the end products of these properties of sedimentary sequences. The responses permit interpretation and prediction of environments of deposition, diagenetic changes, the orientation and trends in sedimentary basins, and in turn, the occurrence of a variety of energy and mineral resources. Thus, from the analysis of vertical sequences, it is possible to understand the nature of changes in shape of the SP and gamma-ray log, which is a critical concern in the operations of the working petroleum and resources geologist. The geometry of sedimentary rocks is the direct product of sediment–fluid interactions, and it controls also the shape of hydrocarbon reservoirs and the nature and discontinuities of seismic

reflectors. Here again, understanding sedimentary processes permits better resolution and interpretations of seismic sections, which are critical for finding reservoirs of oil and natural gas.

The primary porosity and permeability of sandstones are controlled by the processes of deposition; they in turn control the shapes of electric and gamma-ray logs obtained in oil field development. The shapes of these logs tend to fall into four basic groups (Figure 1.1) which include the interdigitate, the blunt-base–blunt-top, the sharp-base–sloping-top and the sloping-base–blunt-top (or funnel-shaped) patterns (Selley, 1976; Taylor, 1977). In well-log analysis, these shapes permit recognition of several possible environments of deposition, and basin mapping from them is possible as shown by several workers (Fisher et al., 1971; Taylor, 1977). However, the shapes are not unique to one specific environment and can be altered by diagenetic changes. Other supporting data must be used. Selley (1976) suggested using combinations of the presence or absence of carbonaceous detritus and glauconite as a means of making further distinctions. Such distinctions can be identified more positively if such data are incorporated with outcrop observations, isopachous mapping, core analysis (see R.R.

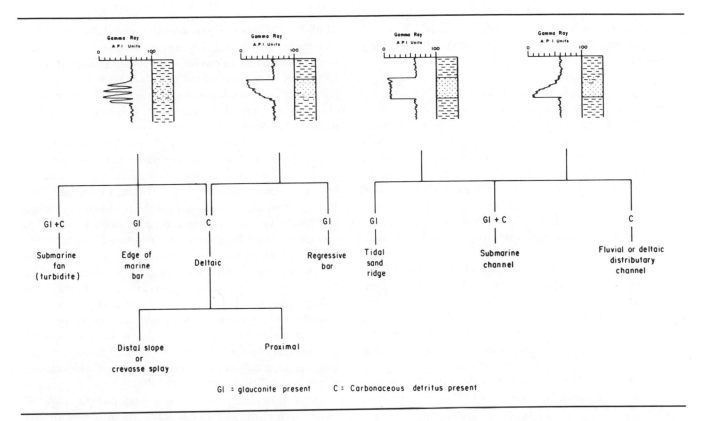

FIGURE 1.1. *Four characteristic gamma-ray log patterns in clastic sedimentary rocks. From left to right: interdigitate pattern of thinly interbedded sandstone and mudstone, an upward-coarsening (funnel-shaped) profile with abrupt top; sandstone with sharp base and sharp top; and a fining-upward sequence with abrupt base. The patterns are not diagnostic by themselves, but when combined with data concerning glauconite content and presence of carbonaceous detritus, the origin of many sandstones may be defined (from Selley, 1976; republished with permission of the American Association of Petroleum Geologists).*

Berg, 1968, 1975, among others), seismic sections, paleontological analysis and so forth. In this monograph, these relations are examined further.

A relatively recent development was the improved understanding of the changes in mineralogy, cementation, porosity and permeability in response to subsurface and near-surface diagenetic processes. These processes include Eh and pH, increasing depth, increasing pressure, temperature, and original sandstone composition (Scholle and Schluger, 1979; McDonald and Surdam, 1984). These changes are critical considerations in making economic decisions for well completion and the development of an oil field. Such changes are summarized in several excellent papers in books by Scholle and Schluger (1979) and McDonald and Surdam (1984).

The emphasis of this monograph is twofold. First, this monograph concentrates on developing the reader's understanding of fifteen types of sandstone depositional systems that are proven to contain reservoirs of petroleum and natural gas, and in some cases contain uranium and coal. These systems are grouped into six general models and consist of:

1. Alluvial fan and fluvial sand-body models,
2. Eolian sand-body models,
3. Coastal sand-body models,
4. Deltaic sand-body models (including fan deltas),

5. Continental shelf sand-body models, and
6. Deep water turbidite and submarine fan sand-body models.

In this monograph, both the geological and geophysical criteria that permit recognition of such sand bodies as potential stratigraphic traps for exploitation and production are stressed. The outline that is followed is relatively simple. Each section begins with a discussion of Holocene sediment depositional processes where such sand bodies are known to occur. Next, the sedimentary criteria and geometry of each of these sand bodies are reviewed from Holocene sediment data. Third, relevant, well-documented examples of ancient counterpart sandstone bodies are presented.

The second emphasis is on the application of basic sedimentological principles to understanding and interpreting subsurface data. Thus, each model closes with a very brief review of case histories of oil, coal, and uranium field examples. In these examples, electric, gamma-ray and sonic velocity logs are illustrated along with observations from cores and seismic sections.

To be valid, each depositional model must satisfy four functions recognized by Walker (1976, p. 22). Each model:

1. "must act as a norm for purposes of comparison,"
2. "must act as a framework and guide for future observations,"
3. "must act as a predictor in new geological situations," and
4. "must act as a basis for hydrodynamic interpretations for the environment or system that it represents."

The six major depositional models discussed in this monograph satisfy Walker's (1976) requirements as facies models, except, perhaps, the submarine fan model.

One of the emerging developments of the past decade is the recognition of a great deal of variability of sedimentary components responding to the same framework and processes of sedimentation. Such variability was recognized first by Galloway (1975) and Coleman (1976, 1980) in studies of deltas where they demonstrated that the changing intensity of sediment yield, wave energy flux, and tidal energy flux would produce different geometries, vertical sequences, and sand-body alignments in deltas.

Miall (1977) reported considerable variability of vertical sequences in braided streams, and most recently, Normark, Barnes, and Coumes (1984) demonstrated major variability in submarine fan morphology, sediment distribution and dispersal, concluding that at the present stage of knowledge, all submarine fan models may well be premature (Barnes and Normark, 1984).

One reason why caution must be extended in dealing with depositional models is the scale of observations and methods of data acquisition used in defining depositional systems, facies, and models (Normark, Piper, and Hess, 1979; Normark, Barnes, and Coumes, 1984). One major problem is lack of a common data set in each study. Observations come from outcrops, bedding features, structures, sequences, morphological units, seismic surveys, or remote sensing, and each encompasses a data set of different scale. Some classes of correlation are beyond biostratigraphic resolution, and reconstructing basin morphology and system morphology from outcrops is barely possible. Surveying methods in Holocene systems, whether submarine fans, deltas, or tidal flats, sample a scale far larger than the average outcrop. Unless the setting is accessible, such as a tidal flat, the resulting survey methods provide gross anatomy and morphology, but cannot define the fine detail observed in a core or outcrop. Perhaps the only method that bridges these different observations is the use of three-dimensional seismic surveys, such as those described by Brown, Dahm, and Graebner (1981). Figure 1.2 illustrates these scalar differences in observation and data acquisition schemes; one of its consequences is that because of the disparate scalar range with which different geologists work, model building becomes a difficult matter. The attempt in this monograph is to integrate these ranges in scale to define standards for petroleum exploration.

Examples of the models reviewed here occur worldwide and are important for hydrocarbon, coal, and uranium exploration. Although the majority of examples discussed are North American, readers of this monograph should be able to find many additional examples in their own areas of exploration and expertise. In this discussion, only previously published examples are included. These published examples should permit all readers active in exploration to apply new and different concepts to the areas where they are now looking for undiscovered oil, gas, coal, uranium, and other mineral deposits.

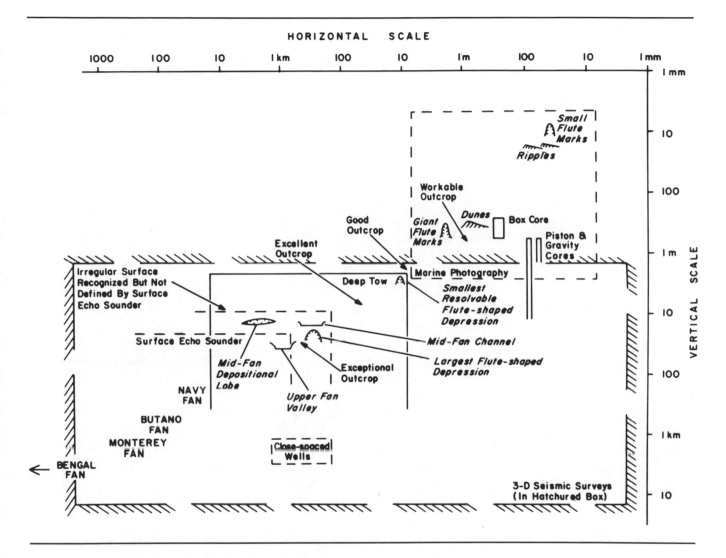

FIGURE 1.2. *Comparison of size of natural outcrops and limits of resolution of standard marine data acquisition and survey techniques with submarine fans and associated sedimentological and morphological features (modified after Normark, Piper, and Hess, 1979).*

It is noteworthy that the field of economic geology is also changing and the process–response approach mentioned earlier is applicable. As ore deposits are being mined out, the exploration strategy is changing from emphasis on "hard-rock-mineralogy-geochemistry" approaches to applications of sedimentology to finding low-grade, economic deposits. Exploration for placer gold is enhanced by knowing the nature of fluvial processes and bar formation (Minter, 1976, 1978; Smith and Minter, 1980). Similar approaches pertain to other heavy metals. Where pertinent, these are incorporated into the text.

Chapter 2

Alluvial Fans and Fluvial Sand Bodies

The sediments of fluvial systems have interested geologists for some time. Considerable literature has grown up around the problems of fluvial processes (Leopold, Wolman, and Miller, 1964; Miall, 1978a), fluvial sediments (Allen, 1965a; Miall, 1978a), and fluvial facies models (Miall, 1978a; Reading, 1978; Collinson and Lewin, 1983). Fluvial sandstone beds are known to contain petroleum reservoirs (Nanz, 1954; Harms, 1966; Berg, 1968; MacKenzie, 1972a; Bloomer, 1977; Campbell, 1976; Putnam, 1982) and also form host rocks for the well-known sandstone-type uranium deposits documented from the Rocky Mountains and the Gulf Coastal Plain (Fischer, 1970, 1974; Galloway, 1977).

Fluvial sediments show the greatest range of textural variability of all the sand-body types reviewed in this monograph. In particular, the sorting of fluvial sediments is variable, most being poorly sorted. Bedding styles and types show remarkable lateral variability over short distances in outcrops and within a subsurface reservoir of fluvial origin. Nevertheless, in certain cases, particularly with fluvial systems on coastal plains, sorting improves and such fluvial sandstones show excellent permeability and porosity, and in subsurface studies are known to contain excellent petroleum reservoirs.

ALLUVIAL FANS

Alluvial fans occur in both arid and humid regions where a sharp reduction in depositional slope occurs and where channel flow changes from confined to unconfined flow. The fan shape and wedgelike third-dimensional geometry

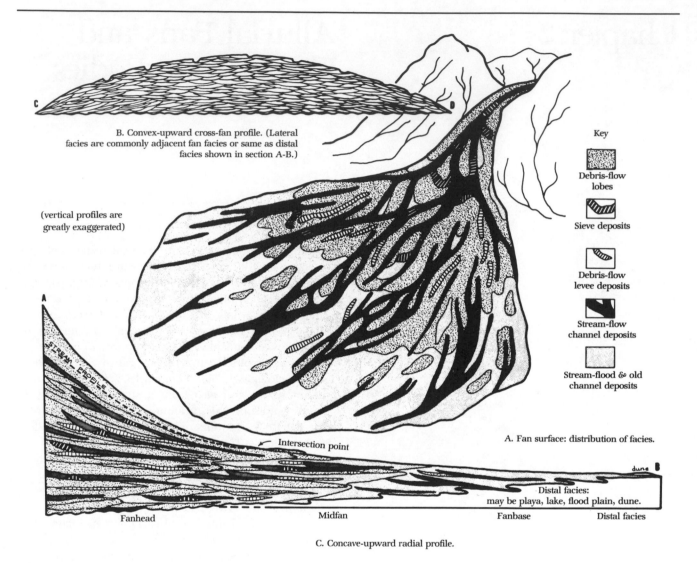

B. Convex-upward cross-fan profile. (Lateral facies are commonly adjacent fan facies or same as distal facies shown in section A-B.)

(vertical profiles are greatly exaggerated)

Key

Debris-flow lobes

Sieve deposits

Debris-flow levee deposits

Stream-flow channel deposits

Stream-flood & old channel deposits

A. Fan surface: distribution of facies.

STREAM-PROFILE

Intersection point

dune

Distal facies: may be playa, lake, flood plain, dune.

Fanhead   Midfan   Fanbase   Distal facies

C. Concave-upward radial profile.

represents a form response to such a slope change from relatively large to relatively small, and from confined to unconfined flow. The best development of alluvial fans occurs, therefore, along fault scarps and fault-line scarps. The alluvial fan form owes its origin to a sudden reduction in open channel flow velocity resulting from both a change to unconfined flow and a sudden reduction in slope angle. Both morphology and sediment distribution are controlled by these slope and flow changes (Figures

FIGURE 2.1. *Distribution of sediment facies and morphological profiles of an ideal alluvial fan (from Spearing, 1975; republished with permission of the Geological Society of America).*

2.1 and 2.2) and the resulting changes in depositional processes and flow velocity. The sediment size, the sediment sorting, and stratification types change from the apical end of the fan to the distal toe of the fan in response to such changes. Shifting drainage patterns permit fans to form a sheetlike bajada marginal to a mountain front (Denny, 1967) with a wedgelike internal geometry. The thickness changes of such a bajada increase toward the mountain front with the resulting isopachous trend being parallel to the depositional strike of the basin.

ALLUVIAL FAN PROCESSES

Several depositional processes occur on alluvial fans. The discussion that follows is after previous work by Blissenbach (1954), Hooke (1967), Denny (1967), Bull (1972), and Spearing (1975). The major processes of sedimentation documented from alluvial fans include debris (or mud) flow, sieve deposition, braided stream channel systems, and reworking by wind systems.

Debris Flow. Debris flow is common to alluvial fans, particularly in arid zones; it occurs also on submarine fans. This process is discussed in Chapter 7, to which the reader is referred for a more detailed statement.

Sieve Deposition. Hooke (1967) demonstrated that on arid alluvial fans, many of the sediment surfaces are porous. Consequently, it is not unusual that some of the water that transports sediment filters through the porous zone and leaves behind a lobe of sediment that is stranded by such infiltration. This sediment lobe is poorly sorted and contains sediment sizes from boulder to clay. The lobes occur as linear sediment bodies (Figures 2.1 and 2.2), confined mostly to the apical fan zone, and are oriented approximately parallel to depositional strike.

Braided Channels. Braided channel systems tend to be superimposed on alluvial fan surfaces in response to the moderate to greater slope angles characteristic of alluvial fans. Because of the rapid change in flow velocity of the channels debouching on the fan, sediment dumping is rapid and water tends to flow around such depositional sites and become arranged into a system of braided streams. These streams bifurcate and build local islands

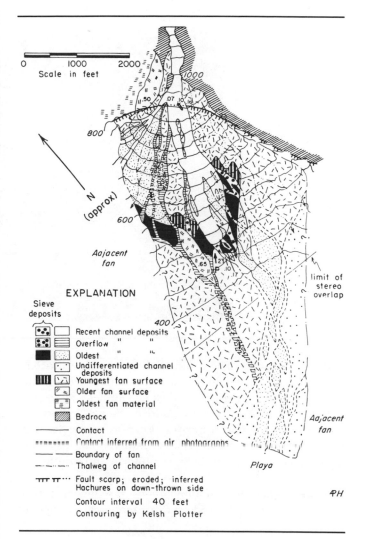

FIGURE 2.2. *Geomorphic map of the Shadow Rock Fan, California (from Hooke, 1967; republished with permission of the University of Chicago Press).*

FIGURE 2.3. *Debris-flow conglomerates showing lack of fabric, poor sorting and lack of stratification in proximal alluvial fan, Mazourka Canyon Alluvial Fan, west side of Inyo Range, California. (Scale is 50 cm long.)*

between channels. The braided channels are character-ized by shallow depths relative to channel width, and new braids are generated as further sediment deposition of bars occurs in response to sudden changes in flow ve-locity. Sorting action improves downstream on the fan and, therefore, the grain size distribution of the distal fan braided stream deposits is more uniform and stratified.

Wind Reworking. On arid alluvial fans, resedimentation of fine- and medium-grained sand by wind systems is com-mon.

ALLUVIAL FAN SEDIMENTS AND FACIES

Sedimentologically, and morphologically, alluvial fans are subdivided into a proximal, medial, and distal facies (Figure 2.1).

Proximal Facies. The proximal alluvial fan facies is char-acterized morphologically by the greatest slope angles on the fan. This facies occurs adjoining the fault scarp or fault-line scarp and comprises the apical zone of the fan complex. The sediment processes in this zone are domi-nantly debris flows or mud flows, with subordinate sieve

deposition and braided channel processes (Hooke, 1967; Bull, 1972). As a consequence, the sediments show a broad range in particle size, are poorly sorted, and lack a fabric. Stratification features are nonexistent. Gravel components are angular. The gravels are grain supported and bound by clay, silt, and sand (Figure 2.3).

Medial Facies. The medial facies occurs in the central por-tion of the alluvial fan and is characterized by braided-channel processes and slightly subordinate debris-flow deposition. Although interbedding of debris-flow and braided-channel deposits is common, some reworking of debris flows results in improved sorting of the sediments. Consequently, sands are interbedded with better-sorted gravels, and the overall sand content also increases. Gravel clasts are imbricated. The interbedded sandstone is parallel laminated (Figure 2.4) and may contain pre-

FIGURE 2.4. *Interbedded gravels of debris-flow origin, gravelly sands of braided-stream origin and coarse sands of braided-stream origin. Sand bed immediately above scale shows buried antidune standing wave bedform with low-angle antidune cross-stratification dipping to right. Mazourka Canyon Alluvial Fan, west side of Inyo Range, California. (Scale is 70 cm long.)*

served antidune bedforms (Figure 2.4) and internal antidune cross-stratification (Figure 2.4) (Hand, Hayes, and Wessell, 1969). Sediment sorting is improved with respect to the proximal facies, but it is still poor. Local thin channel zones with cross-stratified cut-and-fill stratification occur in response to braided stream deposition.

Distal Facies. The distal facies occurs in the toe of an alluvial fan and is characterized by small slope angles. The dominant depositional process in this facies is braided-stream channel deposition. Subordinate wind action may also occur here, with the result that braided-stream deposits and windblown sediments may be interbedded. The distal fan sediments may also intertongue with playa or ephemeral lacustrine sediments in arid settings. The sediment consists dominantly of sand and gravelly sand.

Consequently, sorting is improved, compared with the proximal and medial facies. Parallel laminae, cross-stratification, and imbricated gravel clasts are common to this facies.

ALLUVIAL FAN VERTICAL SEQUENCES

The vertical sequences of alluvial fans tend to coarsen upward overall (Wessell, 1969; Steel, 1976; Steel et al., 1977; Heward, 1978a,b). Fan progradation displaces the proximal facies over medial facies and the medial facies over distal facies. Perhaps one of the best-documented examples of such a history of fan growth and progradation occurs in the Devonian red beds of Hornelen and Solund Basins in Norway (Steel, 1976; Steel et al., 1977). There, several coarsening-upward cycles of fan progradation caused by repeated tectonic uplift have been observed (Figure 2.5) and many of these cycles reflect progradation of several fan facies. Thus, although the sequences are gravel dominated, the proximal facies show coarsening-upward gravel cycles only, whereas the medial facies show coarsening-upward cycles starting with basal sands grading upward into progressively coarser gravels (Figure 2.5).

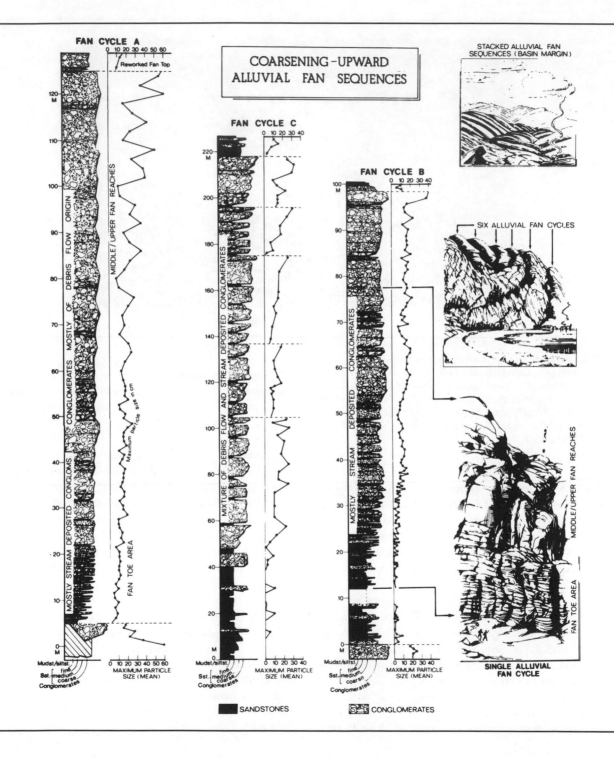

COARSENING-UPWARD ALLUVIAL FAN SEQUENCES

FAN CYCLE A

FAN CYCLE C

FAN CYCLE B

STACKED ALLUVIAL FAN SEQUENCES (BASIN MARGIN)

SIX ALLUVIAL FAN CYCLES

SINGLE ALLUVIAL FAN CYCLE

■ SANDSTONES   CONGLOMERATES

Locally, within individual beds themselves, a coarsening-upward, upper-flow regime vertical sequence can also be observed (Figure 2.6). In the Triassic of Massachusetts, Wessell (1969) documented not only a vertical increase in grain size, but progressive changes from lower-flow regime ripples to plane beds to antidune standing-wave cross-stratification. (Figure 2.6). Antidune standing-wave bedforms (Figure 2.7) are also preserved; these also show internal coarsening (Hand, Hayes, and Wessell, 1969). These local sequences indicate an increase in flow intensity and possibly also of sediment concentration by the transition from braided stream flow to debris-flow processes. A generalized vertical sequence for an alluvial fan is shown in Figure 2.8.

ANCIENT EXAMPLES OF ALLUVIAL FANS

Numerous examples of ancient alluvial fan facies have been documented in the rock record. Some of the best documented examples include the Devonian of Norway (Nilsen, 1969; Steel, 1976; Steel et al., 1977), the Cambrian Van Horn Sandstone of Texas (McGowen and Groat, 1971), the Waterberg Group (Precambrian) of South Africa (Erikkson and Vos, 1970), the Carboniferous of the Maritime Provinces of Canada (Belt, 1968), the Carboniferous of Spain (Heward, 1978a,b), the Triassic of Wales (Bluck, 1965), the Triassic of Nova Scotia (Klein, 1962), the Triassic of the Connecticut Valley (Wessell, 1969; Hand, Hayes, and Wessell, 1969), the Copper Harbor Conglomerate (Precambrian) of Michigan (Elmore, 1984), and the Neogene Violin Breccia of California (Crowell, 1974). In the Triassic of Massachusetts, the Mt. Toby Conglomerate comprises an alluvial fan sequence whose grain-size distribution, paleocurrents, and present-day morphological expression parallels Holocene fans (Wessell, 1969). In the Devonian of Norway (Steel, 1976; Steel et al., 1977), the Carboniferous of Spain (Heward,

FIGURE 2.5. *Internal details of three alluvial fan coarsening-upward sequences, Devonian of Hornelen Basin, Norway. Sketches on right show nature of outcrop expression of such coarsening-upward alluvial fan cycles (from Steel et al., 1977; republished with permission of the Geological Society of America).*

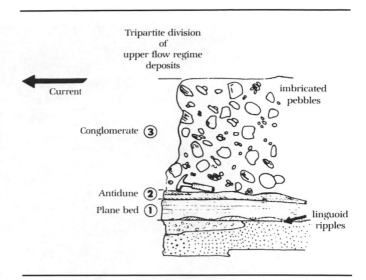

FIGURE 2.6. *Upper flow regime sedimentary sequence in Triassic alluvial fan (from Wessell, 1969; republished with permission of the Dept. of Geology, Univ. of Massachusetts).*

1978a,b), and the Triassic of Massachusetts (Wessell, 1969) coarsening-upward sequences are common both on a local and a bed scale, as well as on a regional scale. These reflect a spectrum of facies (Figure 2.5) organized in a preferred arrangement that one can predict from a normal cycle of fan growth from the abrupt margin of an alluvial basin toward the basin center. Even within a larger complex of alluvial fan deposits, smaller, coarsening-upward cycles occur reflecting the progradation of proximal over medial, and medial over distal zones (Figure 2.5).

The study of Holocene alluvial fans has focused primarily on those in arid-land settings, where details are perhaps better exposed. Nevertheless, it must be emphasized that alluvial fan deposition is in response to basinal slope reduction and change in flow state from confined to unconfined flow, regardless of climatic setting. This point is best illustrated in detailed studies by McGowen and Groat (1971) of the Cambrian Van Horn Sandstone of Texas and by Heward (1978a,b) of some Carboniferous coal deposits in Spain. There (Figure 2.9) alluvial fan

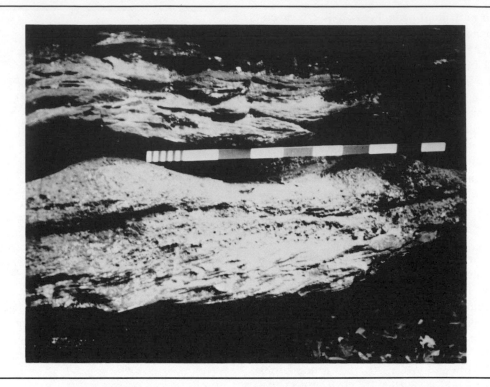

FIGURE 2.7. *Antidune standing wave bedforms with antidune cross-stratification dipping to left, Mt. Toby Conglomerate (Triassic), Sunderland, Massachusetts.*

progradation and growth occurred in a humid setting marginal to a large lake. During periods of channel abandonment, thick coals accumulated on the distal toe of these alluvial fans where the fans intertongued with lake sediments. Coarsening-upward sequences were observed within the fan sequences, showing similarities to the Devonian of Norway.

BRAIDED STREAMS

Braided streams are those that show a bifurcating and anastomosing channel pattern (Figure 2.10); they are common to terrains of moderate slope and moderate to lesser discharge. These streams owe their origins to fluctuations in flow velocity, which cause transported sediment to be deposited as a series of longitudinal braid bars within a channel system (Figure 2.11). Subsequent flow and reworking extends the pattern downstream. This pattern of stream channel flow is characterized also by a

large width-to-depth ratio of the stream (Figure 2.11), a feature favored by abrupt increases in sediment yield to the alluvial plain (N.D. Smith and D.G. Smith, 1984).

The flow patterns of braided streams are unidirectional. The velocity spectrum and flow patterns of braided streams are reviewed by Leopold, Wolman, and Miller (1964), Cant (1978a), Rust (1972, 1978), Williams and Rust (1969), N.D. Smith (1970, 1971), and Miall (1977, 1978b). Doeglas (1962) described the sandy braided Durance River of France, which has served as a reference standard for European sedimentologists, but his presentation lacks details about the exact nature of flow processes and sedimentary responses. More complete studies of braided stream sediment deposition and flow condi-

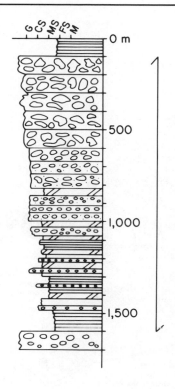

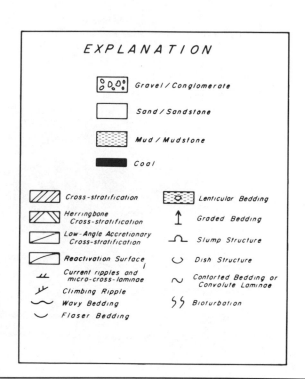

FIGURE 2.8. *Idealized vertical sequence of prograding alluvial fan system. Doubly terminated arrow shows total sequence interval. (Abbreviations: G—Gravel; CS—Coarse sand; MS—Medium sand; FS—Fine sand; M—Mud.)*

tions include those by N.D. Smith (1970, 1971) on the Platte River, by Rust (1972, 1978) and Williams and Rust (1969) on the Donjek River, by Cant (1978a,b) and Cant and Walker (1978) on the Saskatchewan River, by Miall (1977, 1978b) who reviewed the depositional styles of many braided rivers, and by N.D. Smith and D.G. Smith (1984).

BRAIDED-STREAM PROCESSES AND SEDIMENTS
Braided streams tend to form on surfaces of moderate to high slope and to develop longitudinal bars within their channels. Other bar types may occur also and contribute to the braiding of the stream pattern (Figure 2.12). As the bar builds vertically above stream level, or stream level

drops and exposes part of the mid-channel bar, the channel splits and bifurcates. Flow instability caused by such bar development enhances the nonuniform flow velocity, which localizes deposition of additional bars. Bar development causes coarse sediment shields to accumulate on the upcurrent end of individual bars, whereas sand and silt are deposited in the longitudinal direction of the bar (Williams and Rust, 1969; Rust, 1972, 1978; Williams, 1971; Miall, 1977, 1978b; Cant, 1978a). The longitudinal bars are characterized by cross-strata (Williams, 1971; Cant, 1978a,b) and decrease in particle size downstream. Gravelly bars are also common to braided systems, particularly those reported from arctic settings and those adjoining debris-flow deposits of alluvial fans (Rust, 1978).

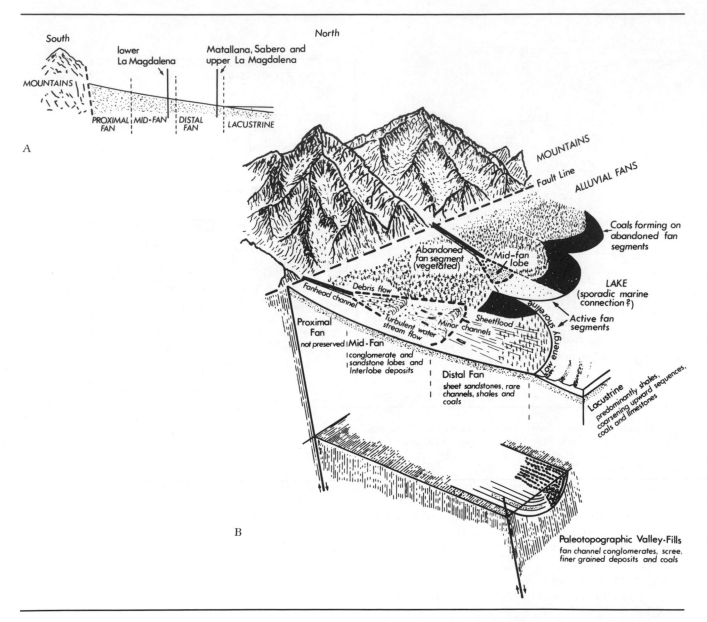

South

North

MOUNTAINS

lower
La Magdalena

Matallana, Sabero and
upper La Magdalena

PROXIMAL
FAN

MID-FAN

DISTAL
FAN

LACUSTRINE

A

MOUNTAINS

Fault Line

ALLUVIAL FANS

Coals forming on
abandoned fan
segments

Abandoned
fan segment
(vegetated)

Mid-fan
lobe

LAKE
(sporadic marine
connection?)

Fanhead channel

Debris flow

turbulent water
stream flow

Minor channels

Sheetflood

Active fan
segments

shoreline

low energy

Proximal
Fan
not preserved

Mid-Fan
conglomerate and
sandstone lobes and
Interlobe deposits

Distal Fan
sheet sandstones, rare
channels, shales and
coals

Lacustrine
predominantly shales,
coarsening upward sequences,
coals and limestones

B

Paleotopographic Valley-Fills
fan channel conglomerates, scree,
finer grained deposits and coals

FIGURE 2.9. (A) *Profile of Stephanian A and B alluvial fan sediments tonguing into lacustrine sediments. La Magdalena, Matallana, and Sabero coal fields, Spain.* (B) *Depositional model of alluvial fans tonguing into subsiding marsh and lacustrine facies, Stephanian A and B (Carboniferous), La Magdalena, Matallana, and Sabero coal fields, Spain (from Heward, 1978a; republished with permission of the International Association of Sedimentologists).*

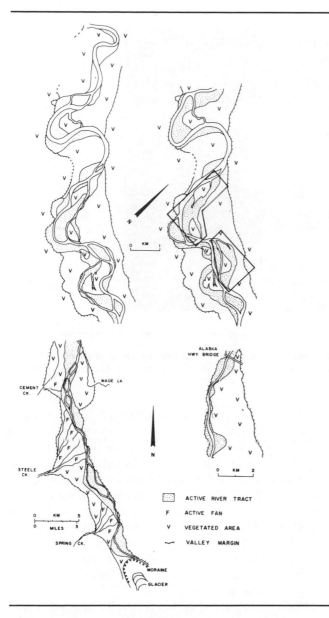

FIGURE 2.10. *Variability of braided drainage patterns, Donjek River Valley, Yukon, Canada (from Rust, 1972; republished with permission of Elsevier Publishing Company).*

Within the braided-stream system, as demonstrated by N.D. Smith (1971) from the Lower Platte River of Nebraska, transverse bars are also known to develop. Their development tends to occur dominantly during cross-flow associated with decreasing discharge following seasonal flooding. Under such conditions, braid bars are cut by newly developed channels that transported sediment into deeper portions of existing channels as a barlike form. Bar progradation in such a localized flow direction maintains and expands the bar. Bar accretion and development are rapid; N.D. Smith (1971) reported the total life span of Lower Platte River bars to be on the order of four to five days. The bars (Figure 2.13) contain parallel laminae, cross-stratification, and ripple bedforms (Rust, 1972, 1978; Cant, 1978a,b; N.D. Smith, 1971; Williams, 1971).

SEDIMENTARY STRUCTURES

The sedimentary structures within braided-stream deposits show great variability of type and represent deposition by both upper-flow regime and lower-flow regime conditions. Parallel laminae are common to most longitudinal bars, as are both avalanche and accretionary cross-stratification. This cross-stratification develops both by bar progradation analogous to the growth of a laboratory delta (see Jopling, 1966) or by normal dune bedform migration within channels and during bar growth (N.D. Smith, 1971; Cant, 1978a,b). Within the channels, the stratification is commonly of the cut-and-fill variety, and includes truncated trough cross-stratification, some with clayey drapes associated with local troughs and swales that are abandoned temporarily. Current ripples normally occur on bar surfaces.

VERTICAL SEQUENCE

The vertical sequences of grain size, lithology, and sedimentary structures reported from braided-stream systems show considerable variability. To the author's knowledge, at least seven types have been reported (Figure 2.14) and both coarsening-upward and fining-upward sequences have been documented (Costello and Walker, 1972; Miall, 1977, 1978b; Cant, 1978b; Cant and Walker, 1978).

In a study of Pleistocene braided-stream deposits, Costello and Walker (1972) reported several coarsening-up-

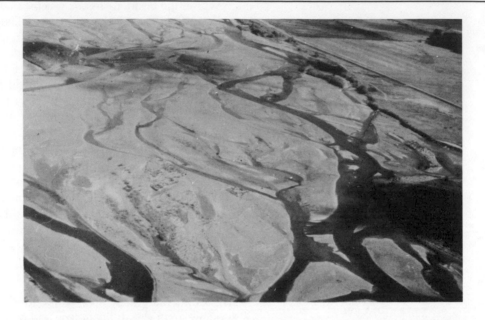

FIGURE 2.11. *Braided-stream pattern with shallow channels and longitudinal bars, Waimakiriri River, Canterbury, New Zealand.*

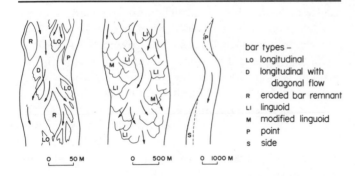

bar types –
LO longitudinal
D longitudinal with diagonal flow
R eroded bar remnant
LI linguoid
M modified linguoid
P point
S side

FIGURE 2.12. *Principal bar types of braided streams (from Miall, 1977; republished with permission of the Elsevier Publishing Co.).*

ward sequences in proglacial outwash braided sandy gravels. The basal portion of the sequence consists of a basal clay representing overbank swale deposition in an abandoned channel; it is overlain by silts and sands with cross-strata, representing deposition during a subsequent flood. As the major channel system shifted into its prior position, coarser, cross-stratified sands and gravels were deposited, after which levee breaching occurred and a succession of thicker sets of cross-stratified sandy gravel was deposited. The sequence shows not only a coarsening-upward grain-size change, but a thickening-upward change in cross-strata sets.

Miall (1977, 1978b) in his review of braided-stream sediments, recognized four sequences (Figure 2.14) from four different Holocene braided-stream systems. These showed overall fining-upward trends in three cases, ranging from gravelly beds to sandy beds, to just sandy braided streams. In addition, he demonstrated that one of the braided streams showed no particular grain-size trend at all. In his later publication, he drew heavily on work by Cant (1978a,b) and Rust (1978) who demon-

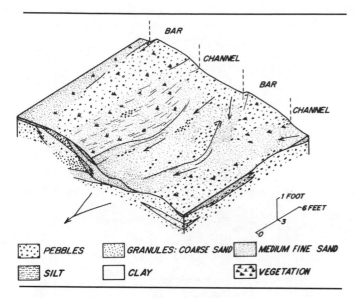

PEBBLES   GRANULES: COARSE SAND   MEDIUM FINE SAND

SILT   CLAY   VEGETATION

FIGURE 2.13. *Three-dimensional model of braided-bar facies (redrawn after Williams and Rust, 1969).*

strated fining-upward sequences for both the sandy braided deposits of the South Saskatchewan River and the gravelly braid deposits interbedded with debris-flow deposits associated with arctic conditions (Trollheim type), respectively. Thus, of the seven types of sequences reported by Miall (1978b) and Costello and Walker (1972), only one coarsens upward. Both Cant (1978a,b) and Rust (1978) reported fining-upward sequences from fossil braided-stream deposits. Figure 2.15 shows a model sequence with subsurface log characteristics.

ANCIENT EXAMPLES

Braided-stream deposits have been identified in many ancient examples. The most convincing cases occur in the Triassic of Nova Scotia (Klein, 1962), the Tuscarora Sandstone of Pennsylvania (N.D. Smith, 1970), the Precambrian Witwatersrand Basin of South Africa (Vos, 1975) where mine maps define braided-stream systems (Figures 2.16 and 2.17), the Devonian Malbaie Formation of Canada (Rust, 1978), the Devonian Battery Point Sandstone of Canada (Cant, 1978b), and braided-stream deposits from alluvial fans (Steel et al., 1977) in the Devonian of Norway. Additional examples were documented by Miall (1977) and by Cotter (1978), who compiled data that suggested that perhaps pre-Silurian fluvial sediments were all braided, whereas post-Silurian fluvial sediments were both meandering and braided, because of the appearance of land plants at that time.

ANASTOMOSING STREAMS

In recent years, a new channel system was recognized by sedimentologists, although it was reported earlier by several geomorphologists (Schumm, 1968). This channel is termed the Anastomosing Channel pattern (D.G. Smith and N.D. Smith, 1980; D.G. Smith, 1983; Smith and Putnam, 1980; Rust, 1981; Rust and Legun, 1983; Putnam, 1982; Putnam and Oliver, 1980). This channel system is characterized by a stable channel pattern that bifurcates and anastomoses into a series of multiple channels (Figure 2.18) and is characterized by small values of sinuosity. Anastomosing channels occur on drainage surfaces of extremely low slope angle (less than 0.5°) in areas that aggrade rapidly with associated large rates of sediment accumulation. Moreover, in order to maintain this channel pattern, large rates of sediment yield from uplifted terrains appear to be mandatory. Concurrent rapid rates of subsidence are required to maintain small depositional gradients (D.G. Smith and N.D. Smith, 1980; Smith and Putnam, 1980; D.G. Smith, 1983). Because of both large sediment accumulation rates and basin subsidence, the preservation potential of anastomosing streams should be large (D.G. Smith, 1983), and perhaps a greater volume of anastomosing channel hydrocarbon reservoirs, such as those described by Smith and Putnam (1980), Putnam and Oliver (1980), and Putnam (1982), exists. These tend to occur in rapidly subsiding intermontane basins or tectonically active foreland plains.

Because two-dimensional channel patterns of anastomosing streams are similar to braided-channel patterns, it is important to review differences between these two channel types. As shown in the previous section, braided channels are characterized by large width–depth ratios and the channels, as a consequence, are extremely unstable. Anastomosing channels, in contrast, show smaller width–depth ratios, and the channels are stable. Consequently, the braided channels are characterized by extensive migration, whereas the anastomosing channel system lacks evidence of migration. That stability appears to be enhanced by extensive zones of wetlands and vegetation associated with humid terrain anastomosing streams (D.G. Smith and N.D. Smith, 1980; D.G. Smith, 1983), al-

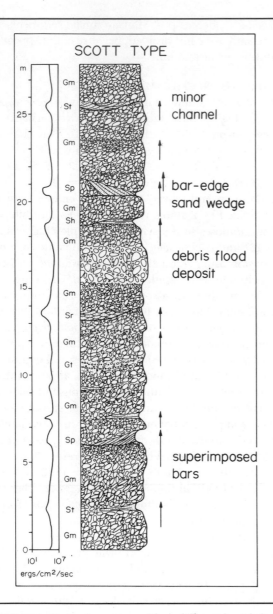

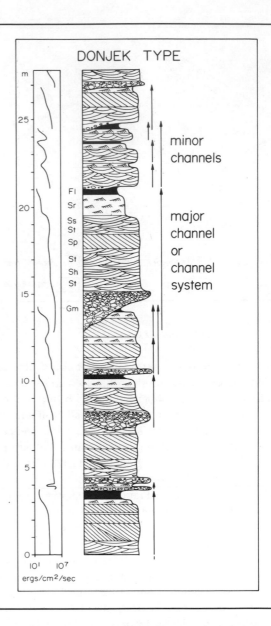

FIGURE 2.14. *Vertical sequence of four different braided-river systems: (from left to right), Scott Glacier outwash river, Alaska; Donjek River, NWT, Canada; Platte River, Nebraska; Bijou Creek, Colorado (from Miall, 1977; republished with permission of Elsevier Publishing Co.).*

though in arid regions (Rust, 1981; Rust and Legun, 1983) the channel stability appears to be enhanced more by large aggradation rates and small gradients. Moreover, as discussed below, anastomosing streams are characterized by a well-ordered three-dimensional arrangement of associated facies, whereas braided channels lack such an ordered arrangement. Anastomosing channels are characterized also by large levees, whereas braided channels contain few or low-relief levees.

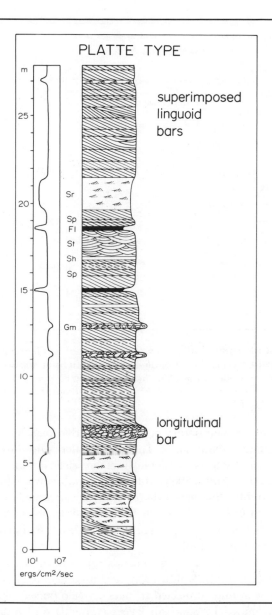

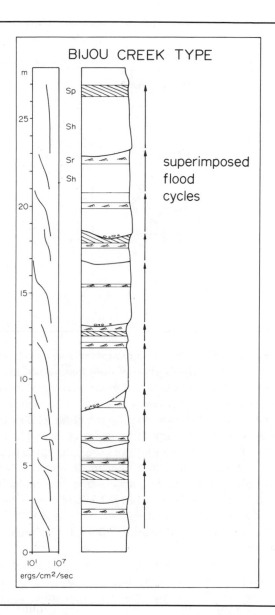

Few data exists concerning the flow velocity characteristics and associated sediments of anastomosing streams. The sediments from the Alexandra River, North Saskatchewan River and Mistaya River near Banff National Park, Alberta (D.G. Smith and N.D. Smith, 1980), the Upper Columbia River of British Columbia and the lower Saskatchewan River of Saskatchewan, Canada (D.G. Smith, 1983) represent the only well-studied anastomosing streams from humid terrains, and Cooper's Creek, of the Lake Eyre Basin of Central Australia (Rust, 1981; Rust and Legun, 1983) represents the only example studied from an arid terrain anastomosing stream. Flow discharge rates were documented only from the Columbia River (D.G. Smith, 1983), where average maximum monthly discharges are reported to be 325 m³/sec and 23 m³/sec for summer high stages and winter low stages, respectively. In contrast, Rust (1981), showed that Cooper's

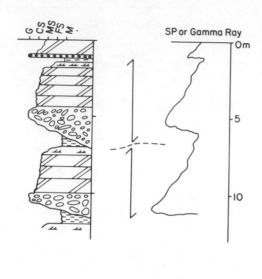

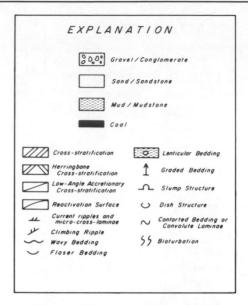

FIGURE 2.15. *Idealized vertical sequence of two braided-stream channel fills. Doubly terminated arrows show thickness interval of each sequence. (Abbreviations: G—Gravel; CS—Coarse sand; MS—Medium sand; FS—Fine sand; M—Mud.)*

Creek, Central Australia, is characterized by intermittent channel flow, with absence of flow lasting several years, because of the arid climate. Flow discharge data during a major flood in 1974 reached 4,000 m³/sec (Rust, 1981) and calculated flow velocities on the channel floor during that flood reached magnitudes ranging from 0.50 to 0.85 m/sec. These flow velocities are large enough to transport gravel ranging from 5 to 10 cm in diameter, a determination consistent with the particle-size distribution found in Cooper's Creek (Rust, 1981). Most of the channels in anastomosing streams are filled with gravel and sand (D.G. Smith and N.D. Smith, 1980; D.G. Smith, 1983; Rust, 1981).

Six separate sedimentary facies are recognized in humid temperature anastomosing streams, and they correspond well with existing environmental subcomponents. These include:

Channels. The channel systems in anastomosing streams are stable and lack evidence of migration. Consequently, they consist of thick to medium-thick fills of gravel and sand, some crudely bedded, some thickly bedded. They extend by avulsion processes through splay cutting into adjoining wetland zones. The channels are confined and stabilized primarily by fine-grained silt and clay com-

prising contemporary or relatively older lake, levee, and vegetated island zones. The channel floors are fashioned into sand waves, particularly where river depths exceed 2 m (D.G. Smith, 1983). Internally, tabular cross-stratified sets, which fine-upward internally, are arranged in multistoried fashion. These sets have been interpreted by D.G. Smith (1983) to represent flood cycles associated with channel aggradation.

The individual channel sands range in thickness from 5 to 15 m (D.G. Smith, 1983) and their geometry consists of linear, interconnected stringers. When buried, they are surrounded by sand-silt levee sediments and mud and peat accumulations (Figure 2.19). The boundaries of these channel sand stringers are sharp and distinct.

Levees. These sediments flank the channel in temperate-humid regions and are prominent because of dense vegetation and development of thick forests. Sediments within the levees consist mostly of silt and organic material; some of this organic material accumulates in distinct

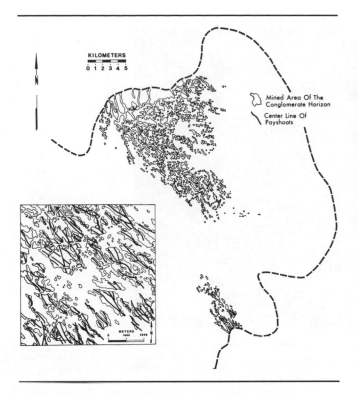

FIGURE 2.16. *East Rand district, South Africa, showing mined conglomerate horizons that define braided-stream pattern and also delineate gold payshoot zones (from Vos, 1975, republished with permission of the Society of Economic Paleontologists and Mineralogists).*

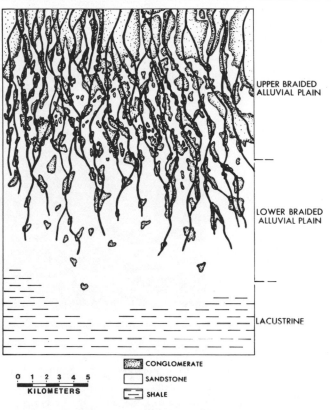

FIGURE 2.17. *East Rand district, South Africa, showing paleogeographic map of Precambrian gold-bearing braided-stream deposits (from Vos, 1975; republished with permission of the Society of Economic Paleontologists and Mineralogists).*

lenses. These silts and clays are internally parallel laminated, but internal rootlet burrowing disrupts this structure. Grain size is observed to decrease laterally within the levees away from the channel margins.

Crevasse Splays. These features are observed to prograde laterally and intertongue with wetland sediments and consist mostly of sand. The splays form as an initial stage of channel avulsion through existing levees, and once channel flow diverges from the main channel through the levee and enters the wetland zone, the flow velocity decreases and a lobate sand splay feature forms. Each of these lobate sheets shows a combination of both a basal coarsening-upward grain size trend from basal silt into plant debris, followed by deposition of sand. As splay sedimentation continues, the channel enlarges in the flow direction and flow velocity and sediment accumulation rates decrease causing deposition of a fining-upward trend ranging from sand to silt and clay. This silt and clay serves as a local substrate for vegetation. Splay size ranges from 0.3 to 1 km² (D.G. Smith, 1983). Internally, laminated silt, sandy current ripples and micro-cross-laminae, and avalanche cross-stratification occur.

Marshes. These adjoin the channel environment and impinge onto levees and comprise part of the overbank spectrum of deposition. The sediments consist of mud interbedded with organic debris; some of this mud is laminated but most is bioturbated, mostly by roots. They

24

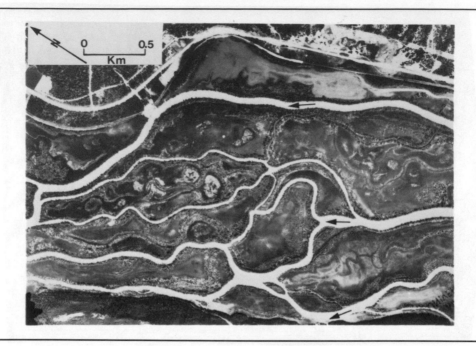

FIGURE 2.18. *Aerial photograph of anastomosing portion of Columbia River, 30 km upstream from Golden, British Columbia, Canada (photograph courtesy of D.G. Smith; from Smith and Putnam, 1980; republished with permission of the Canadian Journal of Earth Science).*

are inundated during flood periods and, because of the small gradient, remain under water much of the time. They occur both on the islands separating the channels as well as the river valley floor.

Peat bogs. These also adjoin the channel environment either on interchannel islands or in overbank zones. They consist mostly of plant debris mixed with mud; the plant debris has undergone decomposition.

Lakes. Irregularly shaped lakes occur in association with anastomosed river systems. They contain laminated silty clay, clay, and minor quantities of organic debris. Bioturbation of sediments is common. These lakes tend to occur at elevations lower than channels and levees and therefore tend to be subjected to avulsion of occasional new channels on a preferred basis. A summary diagram showing the relations of these facies in a temperate-humid zone is shown in Figure 2.19.

In arid regions, many of these subcomponents are preserved (Rust, 1981). Channel systems tend to be longer and deeper and they are flanked by discontinuous levees, with trees developed on them. Side bars of coarse to medium sand occur within the channels, and internally they contain planar cross-stratification and reactivation surfaces. The sediment size tends to fine upward somewhat, particularly in abandoned zones. Deep mudcracks are common along the channel banks. Locally, eolian dunes occur and some of these tongue laterally into the anastomosing channel sediments. Minor volumes of organic accumulations occur.

The major difference between arid and temperate anastomosing channel systems appears to be channel spacing. Nearly 20% of the Cooper's Creek Valley consists of channel facies (Rust, 1981). In comparison with anas-

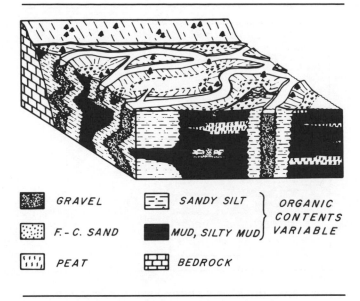

GRAVEL

F. - C. SAND

PEAT

SANDY SILT

MUD, SILTY MUD

BEDROCK

ORGANIC
CONTENTS
VARIABLE

FIGURE 2.19. *Model of predicted textural and bed geometry characteristics of a typical anastomosing river system (redrawn from Smith and Smith, 1980).*

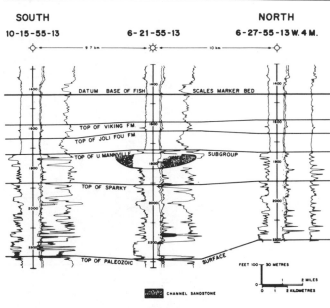

FIGURE 2.20. *Electric log cross-section through typical anastomosing reservoir channel fill, Upper Mannville Group (Cretaceous), eastern Alberta, Canada (redrawn from Putnam and Oliver, 1980).*

tomosing channels near Banff, Alberta (D.G. Smith and N.D. Smith, 1980), this arid channel spacing exceeds humid channel spacing by a factor of 6. As a consequence, the arid anastomosing systems contain less mud and silt than the temperate-humid variety.

VERTICAL SEQUENCE

Data about present-day anastomosing streams (D.G. Smith and N.D. Smith, 1980; Smith and Putnam, 1980; D.G. Smith, 1983; Rust, 1981; Rust and Legun, 1983) are somewhat ambiguous with respect to the type of vertical sequence that is developed in such streams. Because anastomosing streams do not migrate, very little vertical sorting is anticipated, although Rust (1981) suggested that at Cooper's Creek, Australia, a fining-upward trend would develop. In contrast, it could be anticipated that because of lack of migration, no vertical trend in grain size exists. One would expect that in a series of vertically stacked anastomosing streams, one could observe a

coarsening-upward trend reflecting increased relief, discharge, and sediment yield with continued aggradation. Perhaps in this respect, the grain-size trend may mimic the trends observed in an alluvial fan complex.

In ancient counterpart examples, the data are no different. In the Clifton Formation of New Brunswick, a fining-upward trend was reported (Rust and Legun, 1983). In contrast, Smith and Putnam (1980), Putnam and Oliver (1980), and Putnam (1982) displayed some electric log patterns from the Upper Mannville Group (Albian) of Alberta, and the Self-Potential log signature (Figure 2.20) and their textural data suggest that trendless patterns (yielding a blunt-base, blunt-top log pattern) and accessory fining-upward trends (blunt-base, sloping-top log pattern) occur. Better resolution of the vertical sequence of these sediments awaits further work. Nonetheless, on the basis of existing data, a vertical sequence is suggested in Figure 2.21.

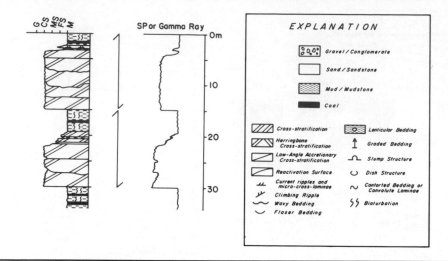

FIGURE 2.21. *Two idealized vertical sequences of anastomosing river systems. Doubly terminated arrows show interval presented by each sequence. (Abbreviations: G—Gravel; CS—Coarse sand; MS—Medium sand; FS—Fine sand; M—Mud.)*

ANCIENT COUNTERPART EXAMPLES

To the author's knowledge, only two stratigraphic units have been interpreted to represent anastomosing river counterparts from comparison with present-day examples. These formations include Member B of the Clifton Formation (Carboniferous) of New Brunswick (Rust and Legun, 1983) and the Albian (Cretaceous) Upper Mannville Group of eastern Alberta (Smith and Putnam, 1980; Putnam and Oliver, 1980; Putnam, 1982). The Clifton Formation is considered a counterpart to the arid-land Cooper's Creek system described by Rust (1981) because of extensive, well-developed channel systems, a fining-upward grain-size trend, the preservation of calcareous nodules interpreted to be caliche, extensive root burrowing, preserved mudcracks, and a consistent paleocurrent orientation trend with little variance, a property considered to be diagnostic of anastomosing channels because of their lack of migration. By contrast, the Upper Mannville Group, described by Putnam (1982), Smith and Putnam (1980), and Putnam and Oliver (1980), is a subsurface unit. Its anastomosing origin was determined from subsurface mapping of channels that showed an anastomosing pattern (Figure 2.22), log shapes showing both a fining-upward trend, or a lack of a trend (Figure 2.20),

and preservation of adjoining channel facies such as levees, crevasse splays, and vegetated, peaty interchannel regions (Figure 2.23). The channel fills consist of sandstones and conglomerates and are hydrocarbon producers (Figure 2.20).

It is noteworthy, however, that Smith and Putnam (1980) compiled some examples of ancient stream deposits that may well be of an anastomosing origin. These examples include the Kootenay-Blairmore and Belly River-Paskapoo Formations (Cretaceous) of British Columbia, the Kootenay Formation (Cretaceous) of Alberta, and the Permian of South Africa. They based their suggestion on the fact that all these examples contain thick freshwater coals requiring overbank zones that were stable over a long time period (approximately 50,000 years), implying delicately balanced water levels, plant growth and peat (and ultimately coal) development. That type of balance suggests a stable channel system, which to them implies an anastomosing channel system. Their point is well taken inasmuch as Ferm and Cavarac (1968) described many cases of channel avulsion being preserved within the Coal Sequences of the Allegheny Group in Pennsylvania and pointed out that their data did not fit too well with a meandering system. It is conceivable that in order to develop and preserve thick nonmarine coal swamp sequences, channel stability is required. Reexamination of such coal measures will determine whether in fact Smith and Putnam (1980) are correct in their reinterpretation.

MEANDERING STREAMS

The meandering alluvial valley system is the best documented, sedimentologically, of the four alluvial models reviewed herein. The data for the meandering alluvial valley model come from studies of the Lower Mississippi River (Fisk, 1944, 1947), the Red River (Harms, MacKenzie, and McCubbin, 1963), the Brazos River (Bernard, LeBlanc, and Major, 1962; Bernard et al., 1970), the River Endrick of Scotland (Bluck, 1971), the Lower Wabash River of Illinois (Jackson, 1975, 1976), the South Esk River of Scotland (Bridge and Jarvis, 1982), and the River Dare of England (Hooke and Harvey, 1983), among others.

FLOW PROCESSES

The hydraulics of meandering flow systems has been reviewed by Leopold, Wolman, and Miller (1964), Allen (1965a), Jackson (1975), and Komar (1900, 1904). The meandering form of the channel superimposes on the flow system a nonuniform velocity pattern (Figure 2.24), in part to minimize drag. Greater velocities of flow are confined to the thalweg, or deeper portion of the channel, and smaller velocities appear to be more characteristic of point bar zones. The flow velocity changes, in a downstream direction, from larger to smaller and back to large as a reference flow line passes through the deeper thalweg, over a point bar and through the next thalweg downstream. As a consequence, the linear path of sediment transport tends to follow the flow lines of stream channel flow, rather than crossing back and forth. This mode of sediment transport was documented experimentally by Friedkin (1945). Thus, point bars build by elongation in a downstream direction as an enlarged longitudinal bar. Point bars may in fact be more complex and

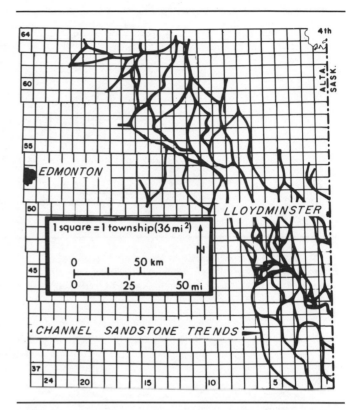

FIGURE 2.22. *Map showing channel pattern of anastomosing channel reservoirs in Cretaceous Upper Mannville Group, eastern Alberta, Canada (redrawn from Smith and Putnam, 1980).*

consist of multiples of longitudinal bars (Figure 2.25), which may be separated by clay (Jackson, 1976). They tend to accrete during bankfull stage (Bridge and Jarvis, 1982). Later cutting by the stream, however, displaces the locus of bar development; thus lateral bar migration involves the accretion of a continuous series of complex longitudinal bars. Channel jumping and splitting is also common to this setting and is accounted for by improved efficiency of stream flow (Fisk, 1947; Allen, 1965a).

The relation of meandering stream velocities to the development of bedforms and to grain-size distributions in the meandering system is extremely complex and involves a fairly close balance of forces and responses to flow by

28

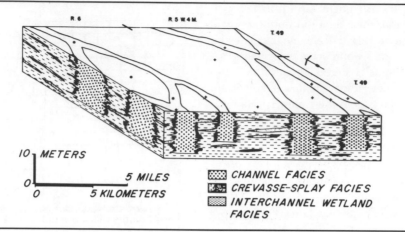

FIGURE 2.23. *Three-dimensional model of anastomosing channel systems and interchannel facies, Upper Mannville Group (Cretaceous), eastern Alberta, Canada (redrawn from Putnam and Oliver, 1980).*

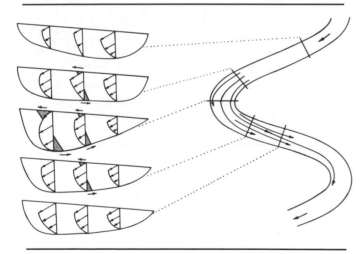

FIGURE 2.24. *Flow pattern and relative velocity magnitude in a meandering river channel (redrawn from Allen, 1965a).*

sediment. These relations have been documented in detail by Jackson (1975) from the Lower Wabash River in Illinois. During a seasonal cycle of flooding, overbank, bankfull and less-than-bankfull flow, considerable changes in flow velocities and associated bedform evolution were observed. Thus, during overbank deposition (Figure 2.26) around the Helm Point Bar of the Wabash River, Jackson (1975) demonstrated that much of the water flow was diverted through a local chute (Blood Neck) and thus downstream from that point flow velocity was considerably less. In response to this hydraulic regime, bedforms (Figure 2.27) were in equilibrium with large velocity flow, but downstream from the Blood Neck the bedforms were lagged (stationary). During slightly less-than-bankfull stage, flow velocities were larger downstream from the Blood Neck (Figure 2.28), and bedforms were migrating actively throughout the meander bend (Figure 2.29). During low water stage, flow velocity was much less, and several areas of lagged (stationary) bedforms were observed (Jackson, 1975).

MEANDERING RIVER SEDIMENTS

The meandering alluvial valley environment consists of several subcomponents or subfacies (Figures 2.30 and 2.31). These include the main channel, which is asymmetric in section, abandoned channels, point bars, levees, and flood plains. Within the channels and point bars, channel flow processes are dominant. Levee aggradation is controlled by overbank flooding and the coarsest sus-

FIGURE 2.25. *Aerial view showing composite longitudinal bars comprising Helm Point Bar, Wabash River, Illinois.*

pended sediment load accumulates there. The flood-plain or backswamp environment is the zone of overbank flooding and the sediment deposited there consists only of suspended load. Abandoned channels form in response to avulsion and channel straightening; once abandoned, they become filled with overbank suspended sediment supplied by overbank flooding.

The sediments and sedimentary facies of each of these subenvironments is fairly well documented. The channel floor tends to consist of the coarsest debris available and consists mostly of gravels and coarse sands, as well as slump blocks derived by undercutting of the cutbank. The channel fill overlying the channel floor consists mostly of sands fashioned into dunes and sand waves; thus internally they are also cross-stratified in relatively thick sets. The lower point bar consists of coarse- to medium-

grained sands that are both cross-stratified and plane-bedded. The upper point-bar environment tends to consist of finer-grained sands with thinner sets of cross-stratification or sets of micro-cross-stratification and a surface of asymmetrical current ripples. The point-bar facies, however, is complicated by local areas of dunes, transverse bars (generating cross-stratification), and longitudinal bars. These longitudinal bars are bounded by clay drapes that separate each bar segment. The grain size changes from relatively coarse to fine across the bar (Figure 2.31) from the lower point bar to the upper zone (Sundborg, 1956), as well as in the downstream direction (Jackson, 1975, 1976). In some instances, coarser sediment distributions change these relations as demonstrated by Bluck (1971) from the River Endrick in Scotland, and by Jackson (1978) from several meandering

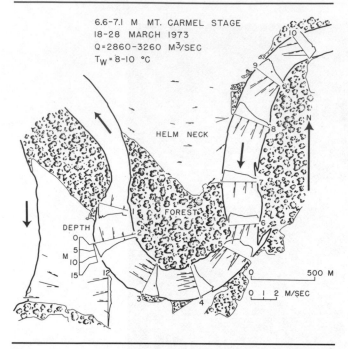

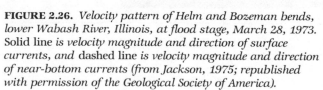

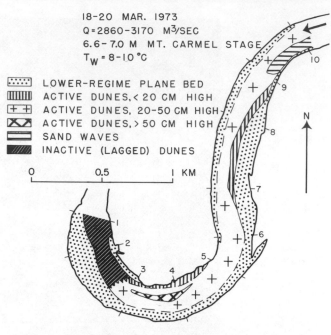

FIGURE 2.26. *Velocity pattern of Helm and Bozeman bends, lower Wabash River, Illinois, at flood stage, March 28, 1973. Solid line is velocity magnitude and direction of surface currents, and dashed line is velocity magnitude and direction of near-bottom currents (from Jackson, 1975; republished with permission of the Geological Society of America).*

FIGURE 2.27. *Distribution of bedforms, Helm and Bozeman bends, flood stage, lower Wabash River, Illinois, 18–20 March 1973 (from Jackson, 1975; republished with permission of the Geological Society of America).*

systems. Muddy meandering systems (Jackson, 1978) also show pronounced differences from the grain-size trends reviewed above.

VERTICAL SEQUENCES

The prevailing paradigm of meandering vertical sequences was that because of the dominant mode of lateral sedimentation in such settings, fining-upward sequences were developed (Nanz, 1954; Bersier, 1958; Bernard, LeBlanc, and Major, 1962; Bernard et al., 1970; Allen, 1963; Visher, 1965, 1972; Sundborg, 1956); a generalized sequence and SP and gamma-ray log patterns are shown in Figure 2.32. This sequence starts with a basal scour overlain by a channel floor lag concentrate of gravel, overlain by coarse sands with cross-stratification of the

channel fill setting. Lying above these strata are the lower point-bar sands, which are medium-grained, and either cross-stratified or parallel bedded. The upper point-bar portion is next above and consists of finer-grained sand, which is micro-cross-laminated. The sequence is capped with the sandy silts, which are characterized by climbing ripple micro-cross-strata of the levee subenvironment and the overbank flood-plain clays and muds.

Jackson (1976) demonstrated that in the lower Wabash River this type of sequence is associated mostly with exposed sections in cutbanks where more of the overbank relations are displayed, but within active point bars, the sequence of structures conforms to the overall model of Bernard, LeBlanc, and Major (1962) except that the grain-size change does not (Figure 2.33). Later, Jackson

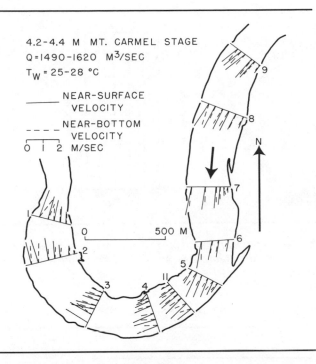

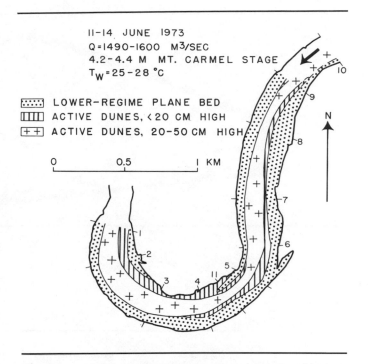

FIGURE 2.28. *Velocity pattern of Helm and Bozeman bends, lower Wabash River, Illinois, at just less than bankfull stage, 11–14 June 1973 (traverses 3–11) and 29 July 1973 (traverses 1 and 2) (from Jackson, 1975; republished with permission of the Geological Society of America). Explanation as in Figure 2–26.*

FIGURE 2.29. *Distribution of bedforms, Helm and Bozeman bends, lower Wabash River, Illinois, at just less than bankfull stage, 11–14 June 1973 (from Jackson, 1975; republished with permission of the Geological Society of America). Explanation as in Figure 2–26.*

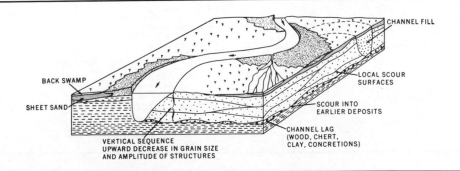

FIGURE 2.30. *Meandering alluvial valley depositional model after Visher (1972; republished with permission of the Society of Economic Paleontologists and Mineralogists).*

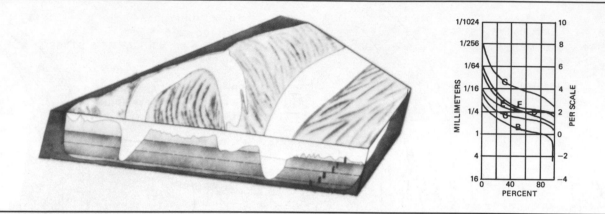

FIGURE 2.31. *Distribution of grain size within sediment sequence in River Klaralven, Sweden (redrawn from Visher, 1965).*

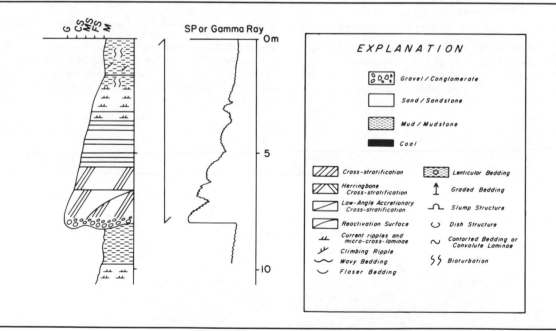

FIGURE 2.32. *Idealized vertical sequence of fining-upward meandering alluvial channel fill. Doubly terminated arrow shows sequence interval. (Abbreviations: G—Gravel; CS—Coarse sand; MS—Medium sand; FS—Fine sand; M—Mud.)*

(1978), in a study of eight small to medium-sized meandering rivers, recognized five lithofacies classes. These are

1. muddy fine-grained streams,
2. sand-bed streams with accessory mud,
3. sand-bed streams without mud,
4. gravelly sand-bed streams, and
5. gravelly streams without sand.

Only the second class shows any close resemblance to a fining-upward model. The remainder do not, and in themselves show great variability. As a consequence, the variability of meandering stream sediments is perhaps as great, if not greater, than the variability of braided stream sediments (Miall, 1977, 1978b). Thus, the *traditional* fining-upward model must be applied with caution, particularly because the upper part of the bars is more difficult to preserve (Campbell, 1976; Plint, 1983).

Nevertheless, it is noteworthy that this fining-upward sequence has been documented from many ancient examples (Figure 2.34) and it in turn controls the characteristics of electric log patterns as shown by Bernard et al. (1970; see also Figure 2.32), R.R. Berg (1968), Harms (1966), and MacKenzie (1972a) among others (see discussion of oil field examples). Thus in some subsurface studies, such sequences have aided in the prediction of occurrences of oil and gas, and in making environmental interpretations.

ANCIENT COUNTERPARTS OF MEANDERING STREAMS

Allen (1965a) and Cotter (1978) compiled many ancient examples of meandering stream deposits and others have been described independently. Allen's (1964, 1965a) classic study of the Devonian Old Red Sandstone of England (Figure 2.34) is one of the better-documented cases. Similar styles of sedimentation were described by Belt (1968) from the Carboniferous of Nova Scotia, and Ferm and Cavaroc (1968) and Horne et al. (1978) from the Carboniferous coal-bearing rocks of West Virginia, among others. Partial sequences have been described and utilized for paleohydraulic interpretation by Cotter (1971) from the Cretaceous of Wyoming. Masters (1967) and Pryor (1961) have described additional Cretaceous meandering stream examples.

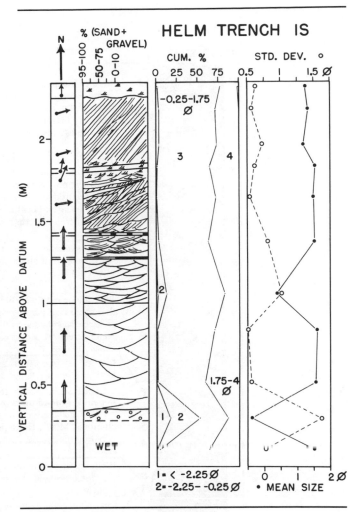

FIGURE 2.33. *Sedimentological log of trench section in lower end of Helm point bar, lower Wabash River (from Jackson, 1976; republished with permission of the Society of Economic Paleontologists and Mineralogists).*

More recently, there have been two outstanding descriptions of fossil meandering stream systems found in Spain (Puigdefabregas and Van Vliet, 1978; Nijman and Puigdefabregas, 1978; Van Der Meulen, 1982; Allen and Matter, 1982). There, unusual exposure conditions permitted three-dimensional observations of outcrops of Eocene, Oligocene, and Miocene sandstones in which details of scrolls bars occur and other sedimentary features

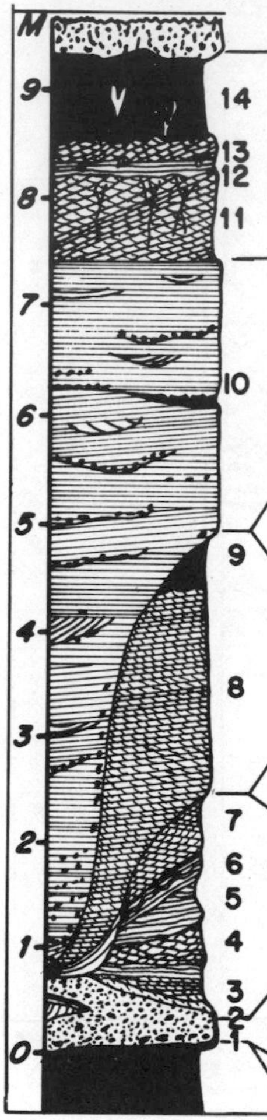

34

Main facts	Interpretation
14 Red, coarse siltstone with invertebrate burrows and abundant calcium carbonate concretions, above **13 12 11** red, very fine, ripple or flat-bedded sandstones with invertebrate burrows.	Vertical accretion deposit from overbank floods. Levee overlain by backswamp deposits. Fluctuating ground-water table during times of exposure.
Fills and covers channel. Red, flat-bedded, fine sandstone with parting lineation, scour and fill, and local scoured surfaces. Scattered **10** siltstone clasts. Local cross-stratification. Lenticular suncracked siltstone.	Channel-fill and lateral accretion deposit. Sand transported as bedload and reworked over shifting channel floor of flat-topped banks. Exposure of higher banks.
9 Cut on siltstone or very fine sandstone in form of channel. Relief about 4.8 m	Erosion at floor of wandering river channel. Extent of wandering possibly controlled by earlier channel-plug
Red, flat or ripple-bedded, very fine sandstone passing up into red, coarse siltstone with carbonate concretions. Scattered siltstone **8** clasts at base.	Channel-fill deposit. Overbank floods plug cut-off channel almost to top. Fluctuating ground-water table and periodic exposure.
7 Cut on very fine sandstones in form of channel. Relief about 2.2 m	Attempt to re-open plugged channel
6 5 4 Repeated graded units overlying scoured surface with small-scale channels. Mostly ripple-bedded, very fine sandstone. Some siltstone.	Channel-fill deposit. Repeated intrusions of suspended sediment down a sloping surface, probably at times of higher stage.
3 2 1 Wedging intraformational conglomerate with sandstone lenticles.	Lag deposit formed at deepest parts of floor of wandering river channel.
Cut on siltstone. Low relief	Erosion of floor of wandering river

FIGURE 2.34. *Fining-upward sequence, Old Red Sandstone (Devonian), Tugford, Anglo-Welsh Borderland, UK (redrawn from Allen, 1965a).*

associated with meandering streams can be observed. These examples are, in the author's view, the best-documented case histories; the reader is referred to the publications cited above for additional details. In another series of regional studies of the Devonian of the Arctic, Friend (1978) documented a general downstream change in sediment features, primarily increases in mudstone and siltstone content that have not been reported from Holocene meandering streams. Once again, studying ancient counterparts has brought out differences in sediment style that have yet to be observed in the Holocene.

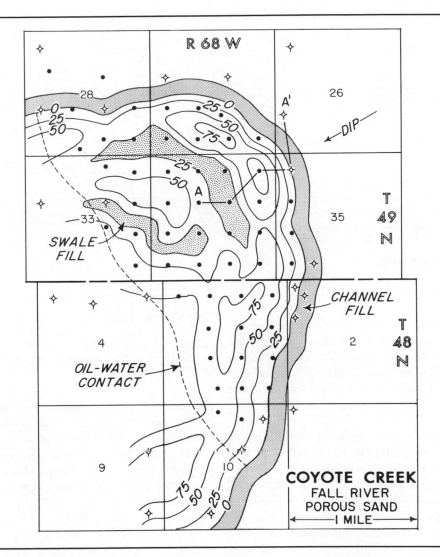

FIGURE 2.35. *Isopach map of Fall River Sandstone point-bar reservoir, Coyote Creek Field, Crook and Weston Counties, WY. CI = 25 ft (from R.R. Berg, 1968; republished with permission of the American Association of Petroleum Geologists).*

OIL, GAS, COAL, AND URANIUM AND GOLD CASE HISTORIES IN FLUVIAL SEDIMENTS

Fluvial sandstones are known to be excellent reservoirs of oil and gas, to be associated with coal deposits, to serve as host rocks for uranium, and to comprise the host rocks for major gold deposits. Examples of these are reviewed below.

OIL AND GAS

To the author's knowledge, the best examples of petroleum and natural gas reservoirs in fluvial sandstones are

36

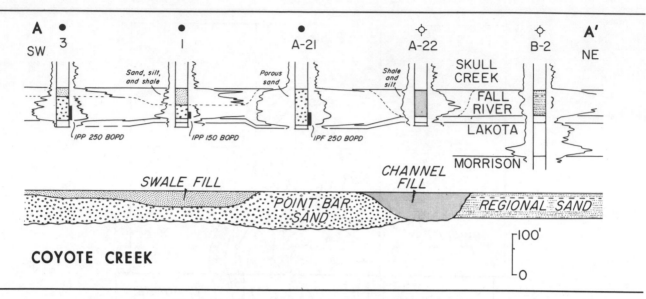

FIGURE 2.36. *Electric log section A–A' of updip side of Coyote Creek field, and interpretation of meander belt zones. Location of section in Figure 2.35 (from R.R. Berg, 1968; republished with permission of the American Association of Petroleum Geologists).*

those that were formed by meandering stream systems. The first such oil field case history to be so documented was the Frio trend (Oligocene) of southern Texas (Nanz, 1954), where meandering alluvial sandstone reservoirs were documented from subsurface features that indicated their organization as a fining-upward sandstone reservoir.

One of the better examples of a fluvial sandstone unit from which oil and gas are recovered occurs in the Fall River Sandstone (Cretaceous) in the Powder River Basin (R.R. Berg, 1968), where eight separate oil fields are of such origin. The evidence R.R. Berg (1968) reported to support his conclusions included the preservation of the point-bar geometry reconstructed from isopachous mapping (Figure 2.35), subsurface mapping of channel margin and channel fill geometries (Figure 2.36), and the nature of the electric log response, particularly the shape of the SP logs (Figure 2.36). These logs show a blunt base, a transitional middle, and a sloping top, which appears to reflect a fining-upward meandering alluvial model. Not all the logs shown in Figure 2.36 reflect this pattern; some show a blunt base and blunt top, for instance, with minor interdigitations within the reservoir sand. Such a log pattern indicates, as Jackson (1978) has demonstrated already, that departures from the fining-upward model occur and caution is required in using upward trend to

identify subsurface reservoirs of meandering point-bar origin. Some of the minor indentations in these logs suggest possible clay drapes within the bar complex itself, clearly indicating that these bars must have formed as a complex system of longitudinal bars as discussed earlier. Such clay drapes would act to provide internal seals for a potential reservoir, and multiple perforations of the well above and below such clay drapes might be required in order to test the reservoir fully.

Harms (1966), MacKenzie (1972a), and Campbell (1976), among others, have reported several case histories of stratigraphic traps from fluvial sandstones in the Jurassic and Cretaceous of the Rocky Mountains. Sinuous channel systems were interpreted from patterns of isopachous mapping, and again fining-upward sequences were recognized from electric logs (Figure 2.37). However, not all the SP logs show a typical fining-upward profile; some show a blunt base and blunt tops with minor interdigitations, again indicating that many of these ancient reservoir sands show significant departures from the fining-upward model, as observed in the Holocene by

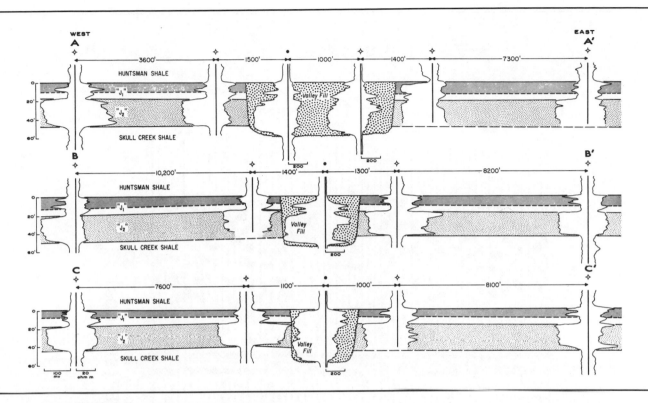

FIGURE 2.37. *Electric log cross-sections showing valley fill stratigraphic trap, "J" Sandstone (Cretaceous), Nebraska. Shape of SP and resistivity curves tends to reflect fining-upward nature of valley fill (from Harms, 1966; republished with permission of the American Association of Petroleum Geologists).*

Jackson (1978). Some of these reservoirs were identified also using field observations of outcrops traced from the subsurface (Campbell, 1976), where partial fluvial sequences were common in multiple-stacked sandstones (see also Bloomer, 1977). Some of the departures from the fining-upward SP or gamma-ray log (Figure 1.1) patterns may be attributed to multiple stacking of partially preserved meandering systems.

Recently, a stratigraphic trap of oil and gas was reported from an anastomosing stream channel reservoir system in the Upper Mannville Group (Cretaceous) of Alberta (Putnam and Oliver, 1980; Putnam, 1982). The channel fills consist of sandstone and conglomerate and were confined by and intertongue with crevasse-splay

silts, and organic-rich mudstone representing stabilized, vegetated islands. Subsurface mapping showed an anastomosing channel system (Figure 2.22) and the channel fills are characterized by a blunt-base, blunt-top pattern (Figure 2.20) consistent with a known lack of grain-size trends in such channel fills. A depositional model for this reservoir is shown in Figure 2.23. Exploration for these reservoirs utilized reflection seismology, and particularly *bright-spot* mapping, to identify so-called low-velocity zones (Figure 2.38).

The seismic character of channel reservoirs was determined (Figure 2.39) from model studies by Harms and Tachneberg (1972) using synthetic modeling techniques, and by Schramm, Dedman, and Lindsey (1977). The main property of fluvial sequences that would be preserved in such seismic records is acoustic impedance. This characteristic requires contrasting acoustic responses between the rocks themselves to produce a section (Figure 2.40), such as developed from refined mod-

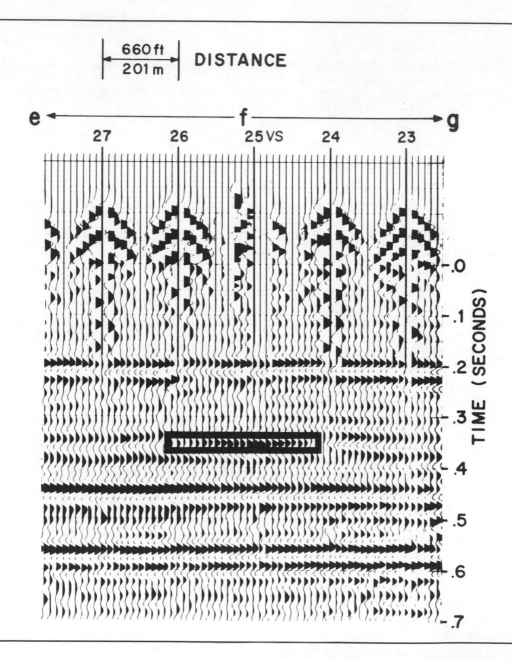

FIGURE 2.38. *Seismic section illustrating bright spot (rectangular box) resulting from gas accumulation in anastomosing channel reservoir, Upper Mannville Group (Cretaceous), eastern Alberta, Canada (from Putnam and Oliver, 1980; republished with permission of the Canadian Bulletin of Petroleum Geology).*

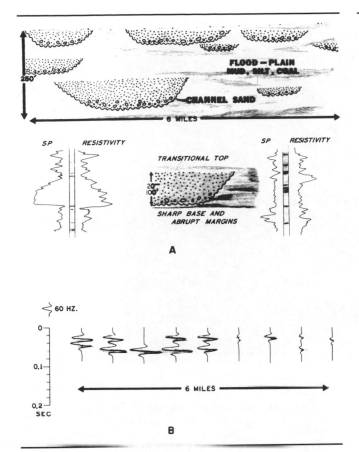

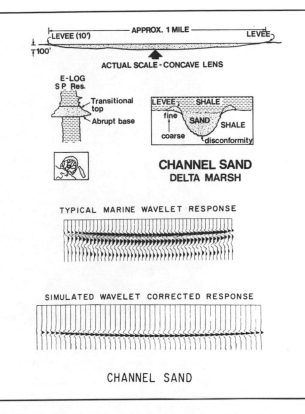

FIGURE 2.39. *Reflectors (seismic) in stream deposits showing (A) distribution of lithologies and control of lithologic and textural properties on electric log characteristics, and (B) a synthetic seismic section of A (from Harms and Tachenberg, 1972; republished with permission of the Society of Exploration Geophysicists).*

FIGURE 2.40. *Comparison of geological and electric log model with typical seismic wavelet and corrected seismic response of a fluvial channel sand (from Schramm, Dedman, and Lindsey, 1977; republished with permission of the American Association of Petroleum Geologists).*

eling techniques by Schramm, Dedman, and Lindsey (1977), among others. Such impedance can be compared with outcrops. Thus, Figure 2.41 shows a channel sandstone cutting down at the expense of underlying mudstones and siltstones. Such a contact would provide acoustic impedance in seismic records that would permit channel recognition. However, Figure 2.42 shows an outcrop of multiple-nested channel sandstones cutting into older channel sandstones. Although these channels appear in an outcrop, the acoustic impedance between these

contacts would be nonexistent, and such sandstones would not be observed in seismic sections at all. In these two cases, the two outcrops shown in Figures 2.41 and 2.42 are within 100 m of each other along a highway and show a change laterally from downcutting into overbank mudstones into multiple-nested sandstones. In a seismic section, partial mapping of the channels would be possible where acoustic impedance would show the channel geometry, but laterally, the geometric resolution would disappear. Thus, the synthetic models of seismic records demonstrate what could be seen in seismic records, provided acoustic impedance and contrast are large enough to bring out these geometric relations. Such is not always the case.

FIGURE 2.41. *View of sandstone channel cutting into interbedded overbank mudstones and siltstones of Caseyville Formation (Pennsylvanian), I-12, 1 mile north of Vienna, Illinois.*

COAL

Coal field occurrences are dominantly associated with delta plain and deltaic systems (see Chapter 5). Nevertheless, in the updip direction from these deltaic zones, coal beds can and do extend into meandering alluvial plains. Several such examples have been documented from the Carboniferous of West Virginia and Kentucky by Ferm (1974), Ferm and Cavaroc (1968), and by Horne et al. (1978) where thick coal seams developed in adjoining overbank flood-plain zones. For thick coals to develop, time is required for thick marshes to develop, usually from rapid accumulation of organic material in combination with similar rates of basin subsidence and compaction. The coal fields of eastern Natal, South Africa (Hobday, 1978) and in the Lower Wasatch (Eocene) and Fort Union (Paleocene) formations of Wyoming (Flores, 1981; Flores et al., 1984; Ethridge, Jackson, and Young-

berg, 1981; Beaumont, 1979) also show similar relationships.

A different type of coal field involving alluvial fan progradation into a lake system was described from the Carboniferous of Spain (Heward, 1978a,b). There (Figure 2.9), coarse-grained, alluvial fan sequences were observed to tongue laterally into lacustrine sedimentary rocks. At the shoreline between the fan and the lake, marshy conditions prevailed. Subsidence rates and compaction rates were equivalent to the rate of peat accumulation yielding thick coals. In this case, a humid alluvial fan system permitted the coexistence of coarse, gravelly beds and extensive peat-forming marshes.

FIGURE 2.42. *View of multiple-nested fluvial channel sandstones in Caseyville Formation (Pennsylvanian), I-24, 1 mile north of Vienna, Illinois. This outcrop is 100 m to the right of the outcrop shown in Figure 2.41.*

Fluvial channel systems will, of course, cut out the lateral extent of coal seams. In the Carboniferous of West Virginia, Horne et al. (1978) discuss means of mapping such channel systems in the subsurface. In the case they presented, a thick coal seam was observed to be incised by a thick channel. In order to decide whether to abandon mining or continue operations, the coal company was advised to drill horizontally through the sandstone to determine its width and paleocurrent trend. These data were then used to demonstrate that by extending operations through the channel, the coal seam could be reentered on the opposite margin, and mining could continue. Even Padgett and Ehrlich (1978) demonstrated that drainage networks of these older channels coincide geographically with present-day surface drainage networks and thus one can predict where ancient channels intersect coal seams by mapping Holocene drainage nets in coal-producing areas.

URANIUM

Uranium ores in sandstones occur mostly in fluvial channel systems. Channel bottom and channel margin settings appear to be the most favorable zones of uranium accumulation (Schlee and Moench, 1961; Eargle, Dickinson, and Davis, 1975; Rackley, 1972, 1975; Fischer, 1970, 1974; Galloway, 1977; Tyler and Ethridge, 1983). In the Gulf Coastal Plain (Figure 2.43), mineralization of uranium is confined to the channels. In other areas, mineralization is favored along channel walls in braided alluvial fans (Figure 2.44), as demonstrated from the Morrison For-

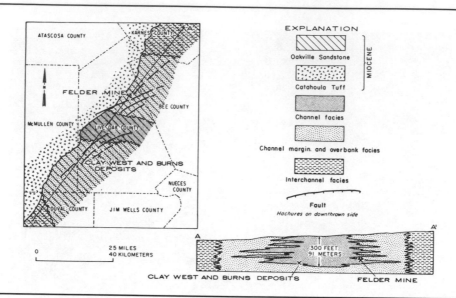

FIGURE 2.43. *Fluvial facies in the Uraniferous Oakville Sandstone (Miocene) of Texas (from Eargle, Dickinson, and Davis, 1975; republished with permission of the American Association of Petroleum Geologists).*

mation by Fischer (1970, 1974) and Rackley (1975), and from the Gas Hills district of Wyoming by Rackley (1972), among other places. Sherborne et al. (1979) demonstrated occurrence of uranium ores in volcaniclastic sandstones and siltstones interbedded with lacustrine limestones and mudstones in Arizona.

The origin of these uranium ores is beyond the scope of this monograph but the following possibilities have been considered.

1. Uranium is coprecipitated by organic-rich lag concentrates in channel bottoms. This relation is observed in the Colorado Plateau (Rackley, 1972, 1975), the Tertiary of the Gulf Coast of Texas (Eargle, Dickinson, and Davis, 1975), and the Permian of Oklahoma (Al-Shaieb et al., 1977).
2. Uranium is transported detritally and concentrated as a placer. This relation is demonstrated in some of the South African examples described by Minter (1976, 1978) and Smith and Minter (1980).

3. Uranium minerals are a diagenetic byproduct of ground-water flow and alteration of unstable iron-bearing minerals derived from granitic and volcanic sources; these uranium minerals are concentrated diagenetically at contacts where ground-water flow is impeded. This process is similar to the *in situ* alteration of iron-bearing minerals that form redbeds (T.R. Walker, 1967, 1975). Several examples of this process have been documented from the Colorado Plateau by Adams et al. (1978) and Tyler and Ethridge (1983), and from the Texas Gulf Coast by Galloway (1977). The ground waters responsible for such a diagenetic accumulation are highly oxidizing and alkaline. Presumably, the precipitation of uranium is accelerated by concentration of organic material along channel floors as suggested above (see also Galloway, 1977).

GOLD

Placer gold mining from stream systems has comprised a major part of the gold extraction industry. Because of its density, gold tends to be concentrated in gravelly channel floors and braid bars. Most of the placer gold mining operations in the Rocky Mountains and California come from such depositional settings simply because the dense gold is hydraulically equivalent to coarse quartz and other light mineral fragments, and accumulates on the lower parts of the channels and side bars (Smith and Minter, 1980).

The Witwatersrand Basin of South Africa is one of the richest areas of gold accumulation in the world. There, gold occurs in Precambrian sandstones that clearly accumulated as paleoplacer deposits in fluvial settings (Pretorius, 1974; Vos, 1975; Minter, 1976, 1978; Smith and Minter, 1980; Buck, 1983; Nami, 1983). Gold is extracted from the floors of such paleochannel fills and lower side bars. Abandoned pay-chutes from which gold has been extracted define the axes of fossil braided channel systems (Vos, 1975; see also Figure 2.16). Within the mine walls, a variety of braid-bar features occur, including lateral bar cross-stratification, variable grain-size changes vertically and laterally, rapid cut-and-fill stratification, and sharp boundaries on basal channels. Preservation of mid-channel bars also occurs (Minter, 1978).

Paleocurrent mapping within the channel systems permits definition of the strategy (Minter, 1976, 1978; Buck, 1983; Nami, 1983) for mapping and predicting new pay zones. The paleocurrent trend defines both the downdip direction of the basin and the axis of channel systems containing gold concentrations. When a particular deposit is mined out, lateral drilling parallel to depositional strike permits one to define additional channel pay-chutes that have not been mapped from vertical drilling from above ground. A cross-cut chute can then be driven parallel to depositional strike, and a new active mine chute can be exploited parallel to the down-channel and downdip direction (Minter, 1976, 1978; Smith and Minter, 1980).

These types of paleoplacer deposits are now recognized elsewhere, including the gold deposits in the Huronian Supergroup (Precambrian) of the Canadian Shield (Mossman and Harron, 1984). There most of the gold accumulations are confined to channel floors of braided stream systems. These deposits are similar to the Witwatersrand stratiform gold deposits, as demonstrated from sedimentary structures, paleocurrent indicators, vertical sequences, and finely comminuted organic material.

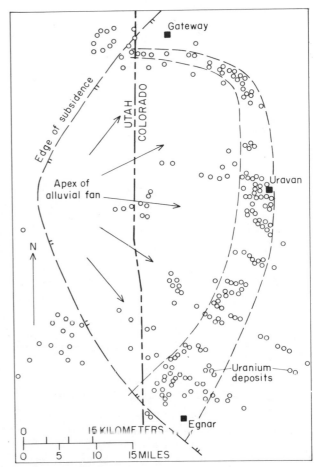

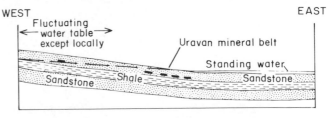

FIGURE 2.44. *Schematic map of Uravan Mineral Belt, Colorado showing outline of paleo-alluvial fan, location of uranium mines, and from them, the general outline of braided-channel systems (from Fischer, 1974; republished with permission of the Society of Economic Geologists).*

Chapter 3

Eolian Sand Bodies

INTRODUCTION

Eolian sand bodies result from sediment transport and deposition by wind systems. Although wind systems operate on a global scale, large eolian sand bodies characterized by an equally large preservation potential, tend to be limited to desert terrains. In humid and polar terrains, moisture adhesion inhibits sediment transport by wind processes. It is only under the dry conditions of arid regions that sediment transport is favored by wind processes. Subsidiary sand bodies deposited by wind are known to occur in coastal regions, and these are discussed in Chapter 4.

SEDIMENTATION PROCESSES

Understanding of sedimentation processes by wind was based on the pioneering work by Bagnold (1941), experimental work by both Bagnold (1941) and Kuenen (1960), and more recent studies by Fryberger (1979a), Ahlbrandt (1979), and Fryberger and Schenk (1981). It was Bagnold (1941) who also first observed the nature of bedform migration in dune areas, but later work by McKee (1966, 1979a,b, 1982), McKee, Douglas, and Rittenhouse (1971), Hunter (1977a,b), Kocurek (1981), Kocurek and Dott (1981), and Kocurek and Fielder (1982), among others, on sedimentary structures in dunes amplified our understanding of these relations.

Overall, the transport and depositional processes of windblown sediments follow the same fluid mechanical principles known from transport of water-laid sediments. The main difference is the effect of transport by a less dense and a less viscous medium. Moreover, wind trans-

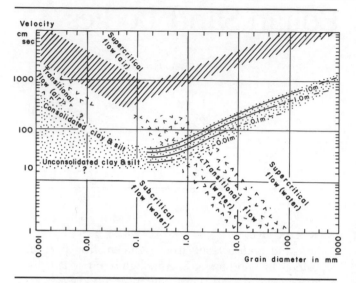

Velocity

Grain diameter in mm

FIGURE 3.1. *Comparison of erosional velocity curves for sediment of differing particle sizes in water (stippled curve) and air (slanted lines) with superimposed flow regimes and zones of transitional flow (redrawn from Sundborg, 1956).*

port involves large-scale unconfined flow. Thus, the erosional velocity of windblown sediments is much greater than for water-laid sediments (Figure 3.1). Generally such transport is restricted to supercritical flow (Sundborg, 1956). The particle size of sediment accumulations is restricted to the sand-sized range, particularly fine- and medium-grained sand, because of their relative lesser erodibility (Figure 3.1). However, it was documented that unusually strong wind storms will transport granules up to 4 mm in size by saltation, and larger particles are known to be transported by wind systems (Sharp, 1979; Sakamoto-Arnold, 1981).

The efficiency of grain transport in wind systems is related to the lesser viscosity and density of air, which permits a more resilient or elastic rebound of grains during transport (Bagnold, 1941; Sharp, 1963, 1964). In water-laid systems, the greater viscosity of water exerts an additional drag on grain surfaces and tends to cushion the impact of grain collisions. In air, this effect is absent, so more resilient rebounding of grains occurs. Thus, the efficiency of sand transport by saltation is increased greatly in wind transport, and the volume of sediment transport

by saltation also appears to increase. The height of the saltation zone above the sediment bed is 1,000 times greater than in water (Bagnold, 1941).

Under wind transport, assuming a mixed grain-size distribution is available, a pronounced separation of processes can be observed. Thus, as the wind velocity increases and sediment erosion begins, most fine-grained silt and clay is dispersed into the atmosphere by suspension transport and is removed almost completely from a given site of sediment transport and deposition (Bagnold, 1941). Sand is transported by saltation (sand driving; Bagnold, 1941) within approximately 0.5 m above the surface and by a surface creep process. The height of saltation will increase with an increase in velocity (Bagnold, 1941; Sakamoto-Arnold, 1981) and the effect of abrasion appears to increase more toward the top of the saltation zone than at the base (Sakamoto-Arnold, 1981). Particle-size zonation occurs within the saltation zone, with coarser material confined to the basal part (Sakamoto-Arnold, 1981).

The sorting characteristics of windblown sand tend to be relatively unimodal with very peaked distributions. Rounded as well as angular grains are common (Folk, 1968, 1978). The grain-surface features tend also to be variable. Some show extensive polishing (Kuenen and Perdok, 1962), whereas other grains show evidence of frosting. However, the frosting appears to be the result of surface chemical reactions (Kuenen and Perdok, 1962) or the overgrowths of very fine-grained quartz terminations that generate a frosted appearance (Amaral and Pryor, 1971). Apparently, frosting of quartz grains is a diagenetic and chemical phenomenon and is not caused by windblown abrasion. Bimodal sands are more common to the interdune or serir zone of desert dunes (Folk, 1968), although they have been observed also in intertidal sand bodies (Klein, 1977a).

EOLIAN DUNES

The end product of wind transport is the development of large asymmetrical forms commonly referred to as eolian dunes. Dune types have been classified according to external morphology (Figure 3.2). The basis of this morphology has been shape, relief, alignment of dune with respect to wind direction and multiple wind directions (Bagnold, 1941; McKee, 1979a,b; Sharp, 1966; Fryberger,

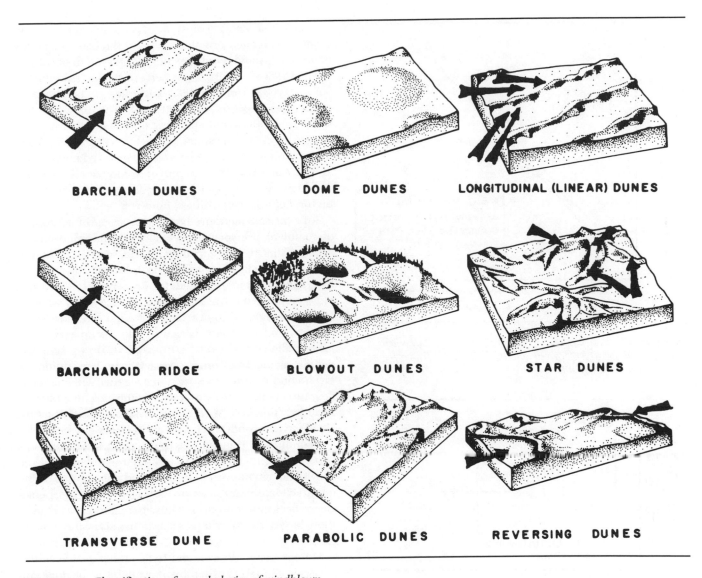

BARCHAN DUNES

DOME DUNES

LONGITUDINAL (LINEAR) DUNES

BARCHANOID RIDGE

BLOWOUT DUNES

STAR DUNES

TRANSVERSE DUNE

PARABOLIC DUNES

REVERSING DUNES

FIGURE 3.2. *Classification of morphologies of windblown dunes in relation to known wind directions, shown by black arrows (redrawn after McKee, 1979a).*

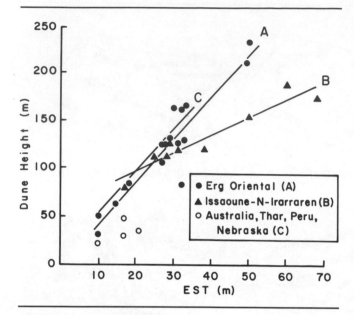

FIGURE 3.3. *Comparison of dune height versus equivalent sand thickness (EST) for various dune fields (redrawn from Wasson and Hyde, 1983).*

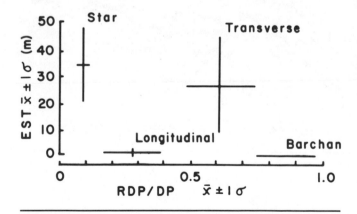

FIGURE 3.4. *Statistically significant separation (P = 0.001) of four major types of dunes according to equivalent sand thickness (EST) and wind directional variability (RDP/DP) (redrawn from Wasson and Hyde, 1983).*

1979a; Tsoar, 1983). Transport of sediment to form these dunes was related primarily to wind direction, changing wind velocity and wind directions, and availability of sediment (McKee, 1979a,b; Fryberger, 1979a; Tsoar, 1983; Sharp, 1966). McKee (1983) suggested a continuum of dune types dependent on sediment availability, starting with dome dunes and changing downwind to transverse dunes, barchan dunes, and ultimately to parabolic dunes under unidirectional flow and greater sediment availability. Andrews (1981) reported a downwind transition from low-relief dunes to intermediate relief barchans to high-relief transverse dunes, however.

An extensive literature has grown around the problem of sediment transport in relation to dune morphology. Much of this literature focused on one, or at the most, two variables. The best summaries on sediment transport and morphology were provided by Fryberger (1979a) and most recently by Wasson and Hyde (1983). Fryberger (1979a) integrated the variables of wind direction, wind velocity, and sediment discharge into a simple ratio related to the Drift Potential (DP) of sand. Drift potential is defined as the sand-moving capacity (discharge) under a given wind regime. Because wind regimes will change in flow direction, the vectoral factor of this potential must be included; this is defined as the Resultant Drift Potential (RDP), where both discharge and directional variability are combined. The ratio, RDP/DP, defines the degree of wind variability under a given discharge regime. Thus, if RDP/DP is small, variability of wind direction is large; unidirectional wind flow favors larger ratios. Morphologies reflect this ratio also. Thus, barchan dunes tend to show largest ratios, whereas star dunes show the smallest ratios.

More recently, Wasson and Hyde (1983) evaluated this parameter in terms of sediment availability for dune migration and demonstrated it to be perhaps the most critical variable involved in dune development and migration. Two parameters were defined for availability of sand, the equivalent sand thickness (EST), defined as the equivalent thickness of sand if dunes were spread out to an even layer, and the dune height. EST represents sediment yield into a dune field, although care must be exercised using it in zones of large wind energy and mobile dune fields. When comparing dune height against EST (Figure 3.3), one can establish a direct correlation, although some variation will occur between different dune fields. When comparing the EST to RDP/DP (Figure 3.4), a distinct separation of four major dune morphologies is demonstrated. Thus, transverse dunes and star dunes require larger sediment yields to form than do longitudinal and barchan dunes. Barchan and longitudinal dunes are distinguished by changes in transport directions, inasmuch as barchans are favored under unidirectional wind flow, whereas longitudinal dunes require two directions of wind flow for their formation (Sharp, 1966; Tsaor, 1983; among others). This multiplicity of wind directions in longitudinal dunes causes a complex system of dune migration involving both downwind transport and reversals, providing net transport in the direction of dominant wind velocity and duration (Sharp, 1966). In contrast, barchan and transverse dunes show a consistent unidirectional migration pattern ranging from 1.0 to 4 m/yr (McKee and Douglas, 1971).

INTERDUNE AREAS

Interdune areas are areas of low relief occurring between individual eolian dunes (Ahlbrandt and Fryberger, 1981; Sharp, 1979; McKee and Moiola, 1975). Such areas fall into three different wind domains reflecting unidirectional air flow (associated with dome, barchan, and transverse dunes), bimodal wind patterns (associated with longitudinal dunes and reversing dunes), and multimodal wind patterns (associated with star dunes) (see Figure 3.2). Sediments accumulating in these areas tend to occur as lenticular bodies buried by dune migration (McKee and Moiola, 1975; Ahlbrandt and Fryberger, 1981).

The major sedimentary processes occurring in interdune areas are either deflationary or depositional, and the presence or absence of water and evaporation processes is significant in characterizing this zone (Ahlbrandt and Fryberger, 1981). Cold-weather processes may also occur with incorporation of snow and later snowmelt-triggered slump features (Steidtmann, 1973).

Dry interdune domains appear to show the best preservation potential. There, wind ripples are most common (see Sharp, 1963), grain-fall strata occur in the lee of dune-slip faces and downwind from occasional vegetation, and sand-flow avalanche accumulations occur also. These are arranged into thin sand sheets, some of which are bioturbated. Scour and fill features are common, particularly around fixed obstructions such as vegetation.

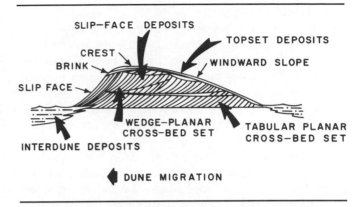

FIGURE 3.5. *Cross-section through typical eolian dune showing terminology for both dune surfaces and internal bedding features (redrawn from Ahlbrandt and Fryberger, 1981).*

Sorting of these sediments tends to be poor; bimodal size distributions are common, in part because of temporary or longer-term deflation processes (Folk, 1968). The relation of interdune sediments to a migrating barchan dune is shown in Figure 3.5.

Wet interdunes differ because the water table is high enough to form temporary or permanent ponds or even lakes. Thus, finer-grained silts and clays tend to accumulate, a larger volume of organic matter may be preserved, and bioturbation features are more common. A greater degree of soft-sediment deformation may occur in these water-saturated sediments once dune deposition occurs directly above. Evaporitic salts may form on drying. Temporary or intermittent water-laid deposits produced by very shallow-water channel flow and lineations and ripples may occur (Sharp, 1979).

Evaporite interdunes differ because they are dominated mostly by chemically precipitated calcite, dolomite, gypsum, or anhydrite. Desiccation features are associated with these sediments; they include traces of coastal sabkhas, particularly in the Persian Gulf (Fryberger, Al-Sar, and Clisham, 1983).

SEDIMENTARY STRUCTURES

Eolian sediments are characterized by a large variety of sedimentary structures, but the most commonly observed type is thick sets of planar and wedge-planar cross-strata (see Figure 3.5 for terminology). These cross-strata contain extremely well-sorted sands, which account in part for their planar and wedge-planar form. Thicknesses are variable, but thicknesses exceeding 10 m are known (Figure 3.6). These thicknesses are caused by the high relief of eolian dunes, but because of either changes in wind directions (such as in longitudinal or star dunes) or changing flow velocities, more complex internal multiple sets of cross-strata (Figure 3.7) have been observed also (McKee, 1966, 1982).

Ripples formed (Figure 3.8) by wind have been described also (Bagnold, 1941; Sharp, 1963). They are distinguished from water-laid ripples by the fact that the coarsest associated sediment accumulates on the crest of the wind-formed ripples, whereas in water-laid ripples the coarsest sediment accumulates in the troughs. Transport of sediment on such ripples involves a combination of saltation and surface creep, with the saltation path defining the ripple wavelength (Bagnold, 1941). Internally, these ripples show very little micro-cross-lamination. Sharp (1963) demonstrated that grain size and velocity are crucial to ripple geometry, particularly because wind-driven sandy ripples show shorter wavelengths, whereas granule gravel ripples show much longer wavelengths.

The internal stratification of windblown sediments is variable. Hunter (1977a,b) reported that plane-bed lamination with internal climbing ripple stratification occurs. The angle of climb of the climbing ripples is variable, because of changing inclinations of the stoss slope. Translacent climbing ripples involve those where no internal micro-cross-laminae are observed and require both subcritical and supercritical flow, leaving a series of low-angle inclined planes capped with a surface ripple form (Hunter, 1977a,b).

A third type of stratification associated with eolian dunes is grain-fall lamination (Hunter, 1977a). These are tabular and bounded sediments that thin as the bounding laminae approach each other and follow topography. Dip angles range from 20° to 28°. This lesser angle permits researchers to distinguish such laminae from avalanche cross-stratification and avalanched sediments.

Sand-flow deposits also occur within large-scale cross-strata of dunes. These are generated by avalanching in fingerlike lobes with an amphitheater-type slump scar at the top (Figure 3.9) and a linear mound downslope and at the base of the flow (McKee, Douglas, and Rittenhouse, 1971; Hunter, 1977a). Internally, these deposits show dis-

rupted layers dipping both up- and downslope interbedded with avalanche cross-strata (McKee, Douglas, and Rittenhouse, 1971). They are cone-, tongue-, or crudely tabular-shaped.

The presence of local moist zones results in the formation of a special type of structure. These are adhesion structures, which are formed by wind transport of dry sand onto wet surfaces; this dry sand adheres to the wet surface and is fashioned into either ripple, wart (or small-scale moundlike), or laminar forms (Kocurek and Fielder, 1982; Hunter, 1977a,b, 1981). Such features were observed first by Van Straaten (1953a) and later by Reineck (1955). Although their origin is linked to wind transport of dry sand onto a wet or damp surface, the features are not environment-specific; they occur on beaches and tidal flats, as well as eolian dunes and interdune regions.

Adhesion ripples (Figure 3.10) tend to form when surfaces emerged from a small pool of water are bombarded by saltating dry sand grains transported by wind. Wrinkle marks serve as nuclei for these ripples, and the structure tends to grow in the upwind direction by addition of dry sand onto wet sand. An inclined, or climbing translation ripple form develops (Kocurek and Fielder, 1982). The adhesion warts were first attributed to shifting wind directions (Reineck, 1955), but later experiments and field observations by Kocurek and Fielder (1982) demonstrated that they form by a vertical grain fall, with local damp nuclei allowing these particles to adhere. The adhesion laminae require the water level to be less than required for adhesion ripples, and level depositional surfaces form as adhesion plane beds. A transition from adhesion ripples to adhesion laminae, reflecting simultaneous sediment accumulation and concurrent reduction of water level, was documented both by Hunter (1981) and Kocurek (1981).

Distinct bounding surfaces are common within sets of eolian cross-strata (Brookfield, 1977; Kocurek, 1981). Several orders of scale of these major bounding surfaces have been identified. The largest, or first order, is nearly

FIGURE 3.7. *Variation in cross-stratification within barchanoid dune, White Sands National Monument, New Mexico. (A) is oriented parallel to dominant wind direction, whereas (B) is oriented normal to dominant wind direction (redrawn from McKee, 1966).*

FIGURE 3.6. *Thick eolian dune cross-stratification, Navajo Sandstone (Jurassic), Zion National Park, Utah.*

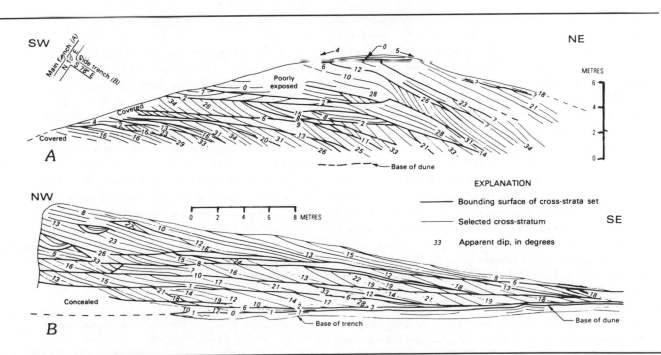

FIGURE 3.8. *Wind ripples, coastal dune, Island Beach State Park, New Jersey.*

horizontal (Figure 3.11) and represents major discontinuities caused by dune migration over large sand seas (ergs) by long-wavelength dunes (draas; Wilson, 1972). As wind strength and sediment yield change, deflation becomes dominant and a base level of migration is developed; usually in the form of a planar interdune surface (Kocurek, 1981). This planar feature is overlain, usually with parallel-bedded sand and overlain by a dune. Second order surfaces develop on the erosional slopes of migrating dunes (Brookfield, 1977; Kocurek, 1981), whereas third order surfaces are reactivation surfaces, representing a temporary truncation. It is the first order surfaces that expose truncated cross-strata representing earlier episodes of dune migration and are distinctive in amalgamated dune sequences.

McKee (1979b, 1982) summarized distinctions in sedimentary structures with respect to dune type. The most common structure in dunes is large-scale cross-strata dipping at a high angle (30° to 34°). These are organized into planar sets or planar-wedge sets. Bounding surfaces between cross-strata are common. In barchan dunes, such cross-strata and internal parallel laminae are diagnostic.

Internally, set boundaries are sharp. The dune exhibits a complex internal organization (McKee, 1966), as shown in Figure 3.6. Transverse dunes also show steep slip faces with internal planar trough cross-stratification, and set boundaries show a curvature parallel to the surface. Parallel laminae of great lateral extent occur parallel to the crest. Parabolic dunes show high-angle cross-strata at the dune base, but these were observed to curve convex upward at a smaller dip angle than in barchan dunes. The dip angles show a spread of orientations of 200°, in contrast to barchan and transverse dunes, which are characterized by a much narrower orientation. Dome dunes are rounded at the top and lack steep avalanche faces on their lee sides. The core shows high-angle avalanche planar cross-strata, particularly on their upwind side, but

FIGURE 3.10. *Adhesion ripples, Cambrian Galesburg Sandstone, The Dells, Wisconsin.*

FIGURE 3.9. *Dune slip face showing wind ripples, avalanche scar, and grain-flow-deposited lobe. Scale is 1 m long. White Sands National Monument, New Mexico.*

FIGURE 3.11. *Eolian cross-stratification and exhumed bounding surface, Permian Lyons Sandstone, Sterling, Colorado.*

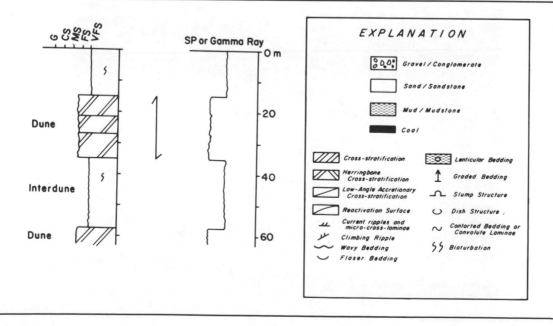

in the upper part the cross-stratification is at a much lower angle (such as 14° to 17°). Locally, cut-and-fill structures are present, but the key indicator is a layer of near horizontal to horizontal layering capping the dune; their orientation boxes the compass and may well be similar to some features described as hummocky cross-stratification (McKee, personal communication, August 1980; Marsaglia and Klein, 1983). Longitudinal dunes contain cross-strata that dip away from the crest because the slopes are basically slip faces, and the dip directions form two clusters nearly 180° apart. Deformation structures have been reported also in dunes (McKee, 1979b) and can be correlated to slumping downslope, or undermining, and grain flowage.

VERTICAL SEQUENCE

Given the complexity of large dune fields (ergs), the limited size distribution of sediments within the dunes, and their complex internal organization, no clearcut vertical sequence has been proposed from actual excavations (McKee, 1966, 1979b). The suggested vertical sequence shown in Figure 3.12 is based on observations by Lupe and Ahlbrandt (1979) from subsurface study of the Pennsylvanian-Permian Tensleep Sandstone, Weber Sandstone and Leo Sandstone of Wyoming, and Clemmensen and Abrahamsen (1983) from outcrop study of the Permian in the Island of Arran, Scotland. In the Permian of Arran, several sequences were identified based on the intertonguing of dune fields with fluvial and alluvial fan facies. In the actual erg facies, the grain size of cross-stratified eolian dunes lacks variation. Only the interbedded granule ripple lag pavements and inverse-graded laminae associated with bounding surfaces showed any change in grain size (compare with Figure 3.7).

ANCIENT COUNTERPARTS

A long list of ancient counterparts of eolian sandstone systems has been documented in the literature (McKee,

FIGURE 3.12. *Idealized Vertical Sequence of interbedded eolian dune and interdune sediments. Doubly terminated arrow shows thickness of dune sequence. (Abbreviations: G—Gravel; CS—Coarse sand; MS—Medium sand; FS—Fine sand; M—Mud.)*

FIGURE 3.13. *Avalanche scar and grain-flow fill on slip face of eolian dune, Permian Lyons Sandstone, Sterling, Colorado.*

1979c). These examples include the Bunter Sandstone (Triassic of Britain; Shotton, 1937), the Keuper Sandstone of England (Laming, 1966; Thompson, 1969), many well-known examples in the Colorado Plateau region of the USA, such as the Coconino Sandstone (Permian) described by McKee (1934, 1945) and Reiche (1938), the Lyons Sandstone (Permian) of Colorado described by Walker and Harms (1972), the Weber Sandstone (Pennsylvanian) described by Fryberger (1979b), the Casper Sandstone (Pennsylvanian) described by Steidtmann (1974), the Tensleep and Leo Sandstones (Pennsylvanian) of Wyoming (Fryberger, Al-Sar, and Clisham, 1983), the DeChelly Sandstone (Permian) of Arizona (Hunter, 1981),

FIGURE 3.14. *Preserved arachnid tracks on serir zone, Permian Lyons Sandstone, Sterling, Colorado.*

the Cedar Mesa Formation (Permian) of Utah (Loope, 1984), and the Navajo, Nugget, and Entrada Sandstones (all Jurassic) described by many workers, but most recently by Hunter (1981). Additional examples are the Galesburg Sandstone (Cambrian) of Wisconsin (Kocurek and Fielder, 1982), a local part of the uppermost Wolfville Formation (upper Triassic) of eastern Canada (Hubert and Mertz, 1980), the Permian of the Island of Arran, Scotland (Clemmensen and Abrahamsen, 1983), and the Permian Rotliegendes of western Europe (Glennie, 1972). All share in common the preservation of thick sets of cross-strata, bimodal deflation reg zones, interdune parallel laminae, major bounding surfaces, preserved avalanche deposits (Figure 3.13) and associated slump scars, wind ripples oriented at right angles to slip faces on which they are exposed, raindrop imprints, adhesion structures, translation ripple features, and tracks of nonmarine animals (Figure 3.14).

The two best-described units, in the author's view, are the Lyons Sandstone of Colorado (Walker and Harms, 1972) and the Permian of Arran (Clemmensen and Abrahamsen, 1983). The Lyons Sandstone is characterized by thick cross-strata organized into planar and planar-wedge sets separated by sharp bounding surfaces (Figure 3.11), preserved serirs with bimodal reg sediments, slump and avalanche scars with avalanche deposits (Figure 3.13), raindrop imprints, superimposed ripples on slip faces, and tracks of nonmarine animals (Figure 3.14). Similar features also are documented in the Permian of Arran by Clemmensen and Abrahamsen (1983) who also demonstrated the importance of intertonguing nonmarine fluvial and fan sequences in their description. As mentioned earlier, they are the only authors to suggest a vertical sequence for dune deposition, consisting of thickly cross-stratified well-sorted fine sand (lacking a vertical

FIGURE 3.15. *Comparison of log shape, dipmeter orientation, and porosity with dune and interdune environments showing direct correlation of subsurface properties with environment in Nugget Sandstone (Triassic) and Sundance Formation (Jurassic), Joyce Creek #1, SE¼, NW¼, Sec 8, T.15N, R103W, Sweetwater County, Wyoming (redrawn from Lupe and Ahlbrandt, 1979).*

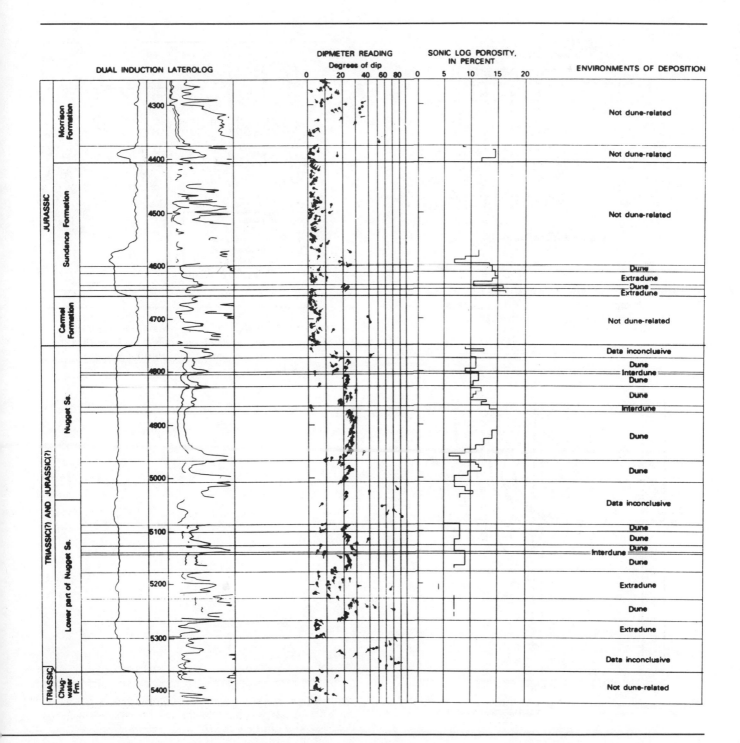

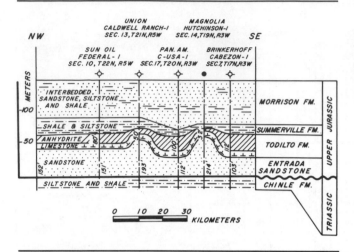

FIGURE 3.16. *Stratigraphic cross-section in Entrada Sandstone (Jurassic) showing preserved dune topography, San Juan Basin, New Mexico (redrawn from Vincelette and Chittum, 1981).*

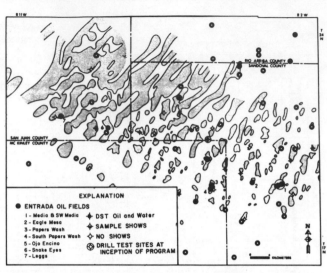

FIGURE 3.18. *Map showing seismic anomalies replicating Entrada Sandstone (Jurassic) eolian dune field, southwest San Juan Basin, New Mexico (redrawn from Vincelette and Chittum, 1981).*

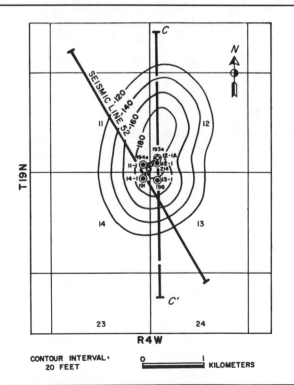

FIGURE 3.17. *Isopach map of Entrada Sandstone (Jurassic), Eagle Mesa Field, San Juan Basin, New Mexico (redrawn from Vincelette and Chittum, 1981).*

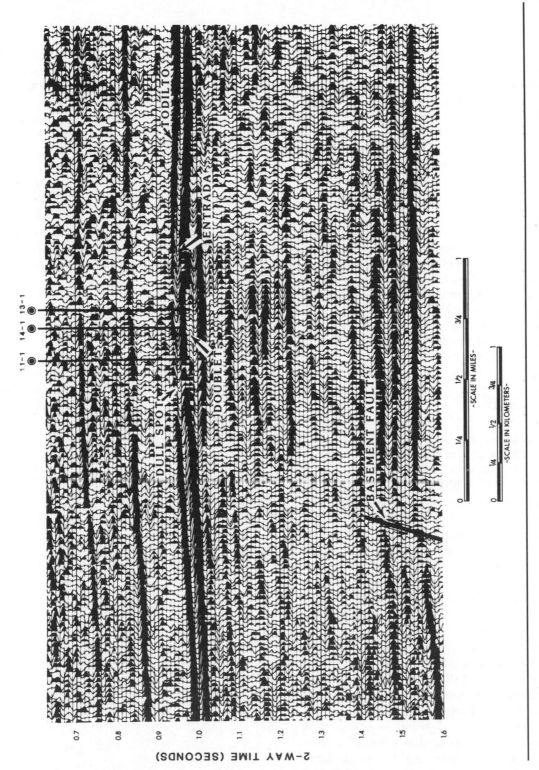

FIGURE 3.19. *Seismic section through Eagle Mesa Field, San Juan Basin, showing doublet signature on Entrada Sandstone eolian dune (from Vincelette and Chittum, 1981; republished with permission of the American Association of Petroleum Geologists).*

grain-size trend), interbedded with interdune bimodal reg deposits and granule ripples, and bounding parallel-laminated sandstone. They documented a large variety of adhesion features, translation ripple bedding, grain-fall lamination and sand-flow cross-bedding. Many of these features were reported in the Mesozoic of the Colorado Plateau by Hunter (1981). It is the association of these features that enables researchers to recognize eolian dune counterparts in the rock record.

OIL FIELD EXAMPLES

Because of the excellent sorting of eolian sands and the size of ergs and sand seas, eolian sand bodies are potential targets for oil drilling. Several examples have been reported, including the Permian Rotliegendes Formation of the North Sea and surrounding regions (Glennie, 1972), the Weber Sandstone at the Rangeley Field in Colorado (Fryberger, 1979b), and both the Entrada Sandstone (Lupe and Ahlbrandt, 1979; Ahlbrandt and Fryberger, 1981; Vincelette and Chittum, 1981) and the Nugget Sandstone (Lupe and Ahlbrandt, 1979; Ahlbrandt and Fryberger, 1981) of the Rocky Mountains and Colorado Plateau of the western USA.

Recognition of these oil field examples as showing an eolian dune origin came from observation of thick cross-stratification in cores, excellent sorting (Glennie, 1972; Fryberger, 1979b; Ahlbrandt and Fryberger, 1981), dip-meter data, and porosity trends (Figure 3.15). In the Nugget Sandstone and Entrada Sandstone (both Jurassic) of Sweetwater County Wyoming, dunes could be distinguished from interdune areas on the basis of dipmeter survey data (Figure 3.15), blunt-base, blunt-top laterolog patterns, and the presence of larger porosity and permeability zones associated with dune sands (Figure 3.15). In the Weber Sandstone at Rangeley Field (Fryberger, 1979b), a similar set of observations identified reservoirs as eolian.

Perhaps the best example of an eolian reservoir system is the Jurassic Entrada Sandstone of the San Juan Basin, New Mexico (Vincelette and Chittum, 1981). The major stratigraphic traps occur on preserved original dune topography (Figure 3.16). These are enclosed by carbonates and evaporites, which tend to thin over preserved dune crests. Isopachous mapping shows not only a closure and a transverse shape (Figure 3.17), but defines the geometry of the reservoir. A systemic seismic survey showed that these eolian dune crest reservoirs were characterized by a specific signature (Figure 3.18) consisting of a doublet shape and a *dull spot* at the top of the dune because of thinning of the overlying limestone. This so-called dull spot represents limited acoustic contrast and the presence of hydrocarbons in the reservoir. By means of continued seismic surveys, a series of identical dune traps was identified (Figure 3.19). Successful drilling on the basis of these surveys has identified many additional producing fields. In this case, by combining knowledge of stratigraphy, the sedimentation model, and seismic surveys, an exploration strategy evolved that permitted new drilling to be successful.

Chapter 4

Coastal Sand Bodies

BARRIER ISLAND SAND BODIES

INTRODUCTION

Beach and barrier island sand bodies occur along and are characteristic of *wave-dominated* coasts of low relief, or coasts that are characterized by sea cliffs of bedrock. The development and preservation of beach and barrier island deposits appear to be controlled by the relative intensity of wave processes and tidal processes along a coastline. As shown by Davies (1964) and later by Hayes (1975), coasts are classified into three groups according to tidal range (Table 4.1). Beaches and barrier islands tend to be preserved best in, and are characteristic of, microtidal and mesotidal coastlines.

Studies dealing with barrier islands and associated beaches share a long history, dating back to DeBeaumont (1845). Some recent studies have provided excellent predictive models for petroleum exploration, including work by Bernard, LeBlanc, and Major (1962) dealing with Galveston Island, Texas, Van Straaten's (1965) study of Dutch coastal barrier islands, Hoyt's (1967), and Oertel and Howard's (1972) studies of barrier processes along the Georgia coast, Hayes' (1969) and Boothroyd's (1978) studies of the New England coast, and work by Hayes (1979), Hayes and Kana (1976), and Barwis (1978) along the South Carolina coast. Kraft's (1971, 1978) study of transgressive shorelines of the Delaware coast is a minor classic and is discussed later. Davis (1985) reviewed most of these studies.

North American sedimentological model building for developing a predictive rationale for petroleum explora-

TABLE 4.1. *Classification of coastlines*

Type	Tidal range
Microtidal	0–2 m
Mesotidal	2–4 m
Macrotidal	> 4 m

SOURCE: Davies, 1964; Hayes, 1975.

TABLE 4.2. *Linear distance of barrier island coasts by continent*

Continent	Linear distance of barrier coasts (mi)
North America	2,000
Europe	500
South America	400
Africa	300
Asia	300
Australia	200

SOURCE: Dickinson, Berryhill, and Holmes, 1972.

tion has, in the view of the author, perhaps overstated the barrier island model. Table 4.2 summarizes the linear distance of the total distribution of barrier coastlines in the world (Dickinson, Berryhill, and Holmes, 1972, p. 193). This table demonstrated that the barrier coasts of the Gulf Coastal Plain and the eastern USA comprise nearly 55% of global barrier island coastlines. This distance encompasses a relatively small percentage of the world's total coastlines. A compilation of tidal ranges on a global scale (Davis, 1964; Hayes, 1975, 1979) demonstrated that well over one-third of the world's present coastlines are macrotidal. Macrotidal coastlines on a global scale may show a preservation potential as great, if not greater than, barrier islands. Because North America contains more atidal coastlines than do other continents, these facts are worth pondering when one considers the preservation potential of barrier islands in the rock record and the potential for finding reservoirs of oil and gas in ancient barrier island counterparts.

DEPOSITIONAL PROCESSES

The major depositional processes operating on barrier islands and beaches are wind-driven wave systems. These wind-driven waves are generated in an open-ocean setting as a result of resistance to wind stress applied to the water's surface. Particles of water in such wind-driven symmetrical waves describe an orbital motion. This orbital motion operates below the water surface to a depth of approximately one-half the wavelength. This limiting depth is known as the null line and is synonymous with the definition of the "surf-base" of Dietz (1963).

Wave systems are driven landward to a shore zone by onshore winds. As these winds approach the shore, the surface base intersects the bottom of the inner continental shelf, and the shape of the wave form changes to asymmetric as the orbital motions of the water particles within the wave become elliptical. The elliptical nature of this movement is a consequence of frictional drag in the lowermost part of the orbital motion, which slows the rate of orbital movement, and the rapid acceleration of the orbital motion of the orbital path at the top of the wave as it accelerates downward. This increased velocity of the orbital motion results in the development of breaking waves and surf (Komar, 1976).

The physical properties of such wave systems are dependent on wind speed and associated wind stress. The wave height, the wave periodicity, wavelength, and depth of sediment scour of waves are directly proportional to wind speed. Consequently, it is during stormy periods that wave action is most effective as an agent of shoreline erosion and sediment transport (Komar, 1976).

The orientation of the approach of wave crests (which define wave fronts) onshore is critical in the development of beaches and barrier islands. Because most wave systems approach shorelines obliquely, the landward portions of wave fronts will shoal earlier than the seaward portions. The effect of such shoaling is to reduce the velocity of the approaching wave front, to reduce the wavelength, and to reorient, or refract, the wave front nearly parallel to shore. Such shoaling processes generate a subsidiary current that is directed parallel to a shoreline in the direction of wave approach. These currents are called *longshore currents;* they transport sediment parallel to a shore and form spits and barrier islands. The velocities and effective lateral extent of such longshore currents are controlled by the height, velocity, and periodicity of the waves, which are in turn governed by wind speed. These variables are also controlled by the area over which wind systems drive water (fetch). Larger longshore current velocities occur during storm periods. During storms, erosion rates of longshore currents increase, the sediment transport volume increases, and consequently major morphological and sedimentological changes occur on barrier islands and on beaches (Figure 4.1). As storms dissipate, rapid sediment deposition and morphological construction occur, repairing much of the morphological damage caused by the storm. Because wind systems change seasonally, the direction of wave approach and resulting longshore current systems also change seasonally (see Fox and Davis, 1978; see also Origin of Barrier Islands).

ORIGIN OF BARRIER ISLANDS

The origin of barrier islands has been a popular topic of research in geomorphology and sedimentology and has generated a large and somewhat controversial literature. DeBeaumont (1845) offered the first proposal to explain barrier island origin, suggesting that barrier islands developed from upbuilding of offshore bars by onshore transport of shoaling waves. He postulated that subsequent accretion of sediment to a height above sea level produced the barrier islands. This model was championed later by Johnson (1919) and evidence in its support was provided in later work by Otvos (1970), on barrier islands along the Gulf Coastal Plain.

G.K. Gilbert (1885) proposed a different process, based on observations made along the coast of Cape Cod, Massachusetts. There, Gilbert (1885) demonstrated that sediment yield by longshore currents formed coastal spits and barrier islands. Onshore transport by waves provided additional sediment. Fisher (1968) showed several examples wherein Gilbert's (1885) model was operative (see also Figure 4.2).

Hoyt (1967) proposed a different model (Figure 4.3) that was related to Pleistocene fluctuations in sea level triggered by the advance and retreat of glaciers. He suggested that eolian ridges, developed on shore during low sea level stands, were submerged partly during the postglacial rise in sea level. The zone behind the submerged ridge became a lagoon, separating the ridge from land and leaving it as a barrier island. Subsequent longshore current sediment transport and wave action resulted in the formation of barrier islands over time.

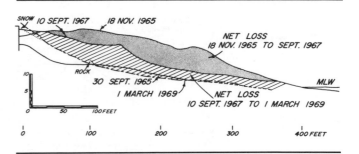

FIGURE 4.1. *Changes in beach profile at Hampton Beach, Hampton, New Hampshire (redrawn from Hayes, 1969). Beach profile of 30 September 1965 was changed by addition of sand by US Army Corps of Engineers to level profiled on 18 November 1965. Depth of storm scour and net loss of sand to September 1967 is shown by* stippled pattern, *whereas additional sand loss to 1 March 1969 is shown by* lined *pattern.*

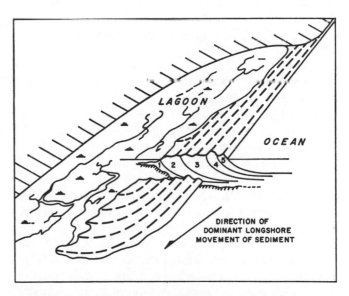

FIGURE 4.2. *Barrier island evolution by combination of longshore current spit formation and periodic breaching by inlet formation (from Fisher, 1968, and Schwartz, 1971; after Gilbert, 1855; republished with permission of the Geological Society of America).*

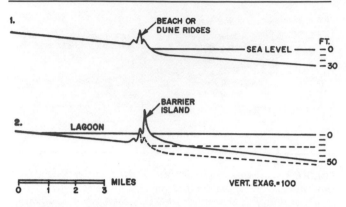

FIGURE 4.3. *Barrier island formation by submergence (from Hoyt, 1967, and Schwartz, 1971). (1) Beach or dune ridge forms adjacent to shoreline, and (2) is submerged by flooding because of sea level rise to form barrier island and lagoon (republished with permission of the Geological Society of America).*

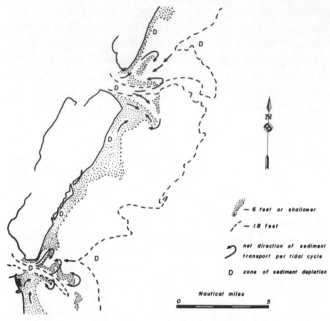

FIGURE 4.4. *Sediment dispersal pattern in zone adjacent to estuarine entrance. Sediment is trapped in closed system at entrance of estuary. Shoal progradation occurs at the expense of adjoining beaches, tidal channels, and shoreface (from Oertel and Howard, 1972; reproduced by permission from Shelf Sediment Transport: Process and Pattern, edited by Donald J. P. Swift et al., Copyright, 1972, by Dowden, Hutchinson and Ross, Inc., Publishers, Stroudsburg, PA).*

Working along the Gulf Coastal Plain, Shepard (1963) stressed the importance of longshore currents in providing sediment, and onshore sand movement by shoaling waves in establishing barrier islands, particularly in connection with transgressive events associated with the destructive history of a delta. This model is, in effect, a composite of both the DeBeaumont and Gilbert models. In short, barrier islands appear to be characterized by a history of multiple causality, as stressed by Schwartz (1971). Local hydrographic conditions apparently control which of these processes are dominant.

Later, a different model was developed from a detailed study of the Georgia coast by Oertel (1972) and Oertel and Howard (1972). Their model is closely dependent upon the presence of estuarine systems along the shore. These estuaries are flushed out during ebb tide by jet flow, which builds up a series of ramp shoals on the seaward side of tidal inlets segmenting barrier islands (Figure 4.4). Breaking waves redistribute such sands from the ramp shoals onto the seaward zone of barrier islands, where longshore current systems provide sand to the barrier islands themselves, causing barrier accretion. During periods when sediment discharge from the estuaries is small, however, the ramp shoals undergo erosion and are destroyed. Then the longshore current systems begin to erode the barrier islands. Barrier island accretion and construction are favored during periods when the ramp shoals are growing, whereas erosion is favored when ramp shoals are cut off from their sediment sources (Figure 4.5). Superimposed on this complex system are seasonal changes controlled by the direction of onshore storms (Figure 4.5). Thus, during certain periods of the year barrier islands undergo accretion and at other times they undergo erosion (Figure 4.5).

SUBENVIRONMENTS OF THE BARRIER ISLAND SYSTEM AND THEIR SEDIMENTS

Barrier islands are divided into a series of subenvironments. From the seaward side to the landward side, these

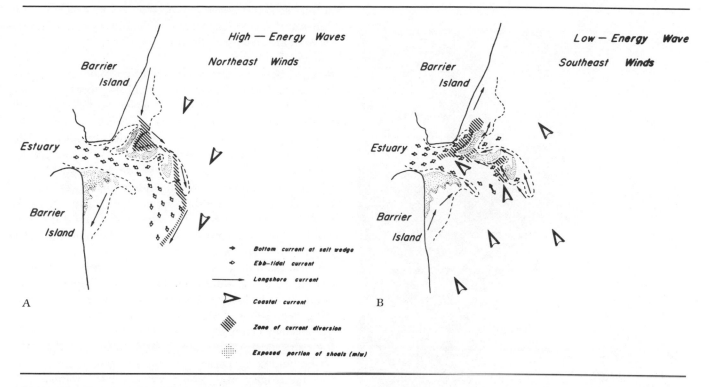

FIGURE 4.5. *(A) Patterns of ebb-tidal currents, longshore currents, coastal currents, and wave systems on Georgia estuarine entrance. Sand shifts from shoals to barriers, and accretion occurs. (B) Low-energy pattern of ebb-tidal currents, longshore currents, and coastal currents, giving rise to shift in barrier buildup, and re-erosion of barriers fed during northeastern storms as shown in (A) (from Oertel and Howard, 1972; reproduced by permission from Shelf Sediment Transport: Process and Pattern, edited by Donald J. P. Swift et al., Copyright, 1972, by Dowden, Hutchinson and Ross, Inc., Publishers, Stroudsburg, PA).*

subdivisions are the foreshore beach, the backshore beach, the coastal dune, the lagoonal marsh, and the lagoonal tidal flat (Figure 4.6). These barriers are cut by tidal inlets, which may be associated with tidal flats.

Lagoonal Flat. The lagoonal flat on the landward side of a barrier island faces and slopes into the lagoonal water mass and is developed where there may be a microtidal to mesotidal tidal range (up to 4 m). These flats are tidal flats in essence and are composed dominantly of mud. A

minor amount of textural zonation occurs on such flats. These flats are laminated and bioturbated.

Lagoonal Marsh. The lagoonal marsh also occurs on the landward side of barrier islands and faces toward the lagoonal water masses. This zone consists of *Spartina* grass, or mangroves that trap muds and silts. The substrate of these marshes usually consists of the muds and silts of the lagoonal flats. During periods of storm activity, washover fans consisting of cross-stratified sands develop in these areas.

Coastal Dune. The coastal dunes are formed on the crest and flanks of barrier islands. The topography of these areas is governed by wind deposition and transport of sediment. Well-sorted, fine-grained sands, organized into wedge-shaped sets of avalanche cross-stratification, characterize this zone. The serirs or troughs of these dunes may show a bimodal size distribution, and they may be covered by wind-generated current ripples. Certain types

FIGURE 4.6. *Aerial view of barrier island, Island Beach State Park, New Jersey, with view to south. Atlantic Ocean to left, and Barnegat Lagoon to right. Washed sandy zone (left) is beach foreshore, whereas backshore is represented by white area to right (west) of washed zone. Coastal dune is in central part with partial vegetation cover. Coastal marsh along Barnegat Lagoon shore.*

of long-rooted grasses may become established on these dunes and develop preserved rootlet burrows that destroy some of the cross-stratification.

Beaches. Beaches are subdivided morphologically into backshore and foreshore zones, with the foreshore sloping seaward. The boundary between the foreshore and the backshore is known as the berm and it marks the farthest point of landward erosion during storms. The foreshore zone is one of sediment accretion produced by the onshore swashing and offshore backswashing of waves. These processes develop a low-angle accretionary cross-stratification that dips seaward at an average angle of 8° to 10° (Figure 4.7). This type of accretionary cross-stratification is diagnostic only of foreshore beach facies in microtidal coasts.

On mesotidal coastlines, the foreshore beach zone changes to a ridge-and-runnel topography. Here, a series of asymmetrical ridges are oriented parallel to the beach with steep-dipping slip faces oriented landward. The troughs between these ridges are defined as runnels. During rising tide, wave action swashes over the ridges, directed landward. These water masses transport sediment over the ridge and deposit it as avalanche cross-stratification on the landward side of the ridge. With time, a set of landward-dipping avalanche cross-stratification is developed (Davis et al., 1972). These may be overlain by the coastal dune sands, which show variable onshore- or offshore-oriented cross-strata also.

Other sedimentary structures common to beaches on barrier islands or beaches *per se* are rhombic ripples, scour marks, swash marks, antidune cross-stratification, cusps, shell pavements, and burrows.

The backshore zone is an area of sediment deposition, and the so-called storm beach zone develops in the zone adjoining the coastal dune complexes. The deposition of coarse sediment is common during storms, as is known from New England pocket beaches (Hayes, 1969). Granule gravel, shell fragments, and environmental debris such as construction timber and refrigerator doors are deposited on wave-cut platforms.

Longer-term storm effects have also been observed. For example, on the coast of Oregon, the summer season is characterized by upwelling and low waves, and net deposition prevails. During the winter season, when winds

FIGURE 4.7. *Low-angle accretionary cross-stratification in foreshore zone at Newport Beach, Newport, Oregon. Scale in cm and 5-cm intervals.*

shift from north to southwest, sand is removed from beaches by longshore currents. Between storms, some of this sand is returned to the beach in the form of longshore bars (Fox and Davis, 1978). The sediment features, however, indicate wave deposition, rather than storm scour.

Tidal Inlet and Tidal Delta. These features are treated together because, in terms of depositional processes, their origins are intertwined. The tidal inlet environment consists of a series of passes through the barrier islands, connecting the lagoonal area with the open ocean. Tidal deltas are fan-shaped bodies, composed mostly of sand, which form either in the lagoon or on the seaward side of tidal inlets and are headed into the tidal inlet (see Finley, 1978, and Boothroyd, 1978, for detailed discussions of the processes and sediments).

The nature of tidal inlet sediments and associated tidal deltas is dependent on the relative interplay of wave-dominated and tide-dominated currents (Hubbard, Oertel, and Nummedal, 1979). Tide-dominated inlets are characterized by relatively deep channels with ebb-dominant circulation and lateral channel margin bars. Flood-tidal deltas are nonexistent or poorly developed in these environments. Wave-dominated inlets are characterized by large flood-tidal deltas building into lagoons; the inlet channels are generally shallow, bifurcating landward and seaward of the inlet threat. Ebb-tidal deltas are small and extend seaward only for short distances in such settings. Transitional situations are also known where sand bodies build up and are concentrated within the inlet *per se.* Wave-dominated inlets are common along the coast of New Jersey, whereas both wave- and tide-dominated inlets and transitional inlets are known from the southeastern USA (Hubbard, Oertel, and Nummedal, 1979). Such

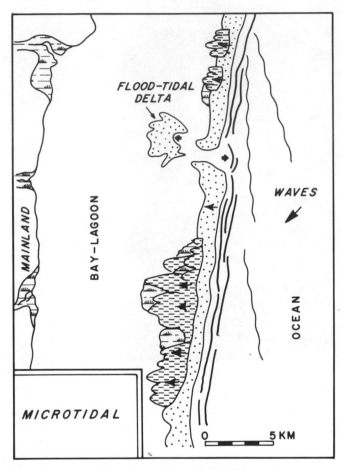

FLOOD-TIDAL DELTA

MAINLAND

BAY-LAGOON

MICROTIDAL

WAVES

OCEAN

0   5 KM

FIGURE 4.8. *Morphology of a typical microtidal barrier island system with medium wave energy showing paucity of tidal inlets and presence of large areas of washover features (redrawn from Hayes, 1979).*

inlet-type sediments are overlain eventually by spit platform and spit sediments (Kumar and Sanders, 1974).

RELATIVE ROLE OF INTERPLAY OF WAVE AND TIDAL PROCESSES ON BARRIER ISLAND MORPHOLOGY AND SEDIMENTOLOGY

This discussion of barrier island features has focused primarily on wave systems, yet as Davies (1964) and Hayes (1975, 1979) have emphasized, coasts may be classified according to tidal range (Table 4.1). Consequently, the relative proportions of wave energy and tidal current processes will shift as tidal range is increased; both mor-

phology and associated sedimentology will change also (Hayes, 1979).

Microtidal Coasts. Microtidal coasts are characterized by varying levels of wave energy, but the tidal range is less than 2 m. Sediment transport in such coasts is clearly dominated by a combination of wave action and longshore currents. The resulting morphology is characterized by long linear barrier islands that are cut by few tidal inlets (Figure 4.8). Intense storm activity favors the development of washover fans. Tidal currents within tidal inlets tend to be flood dominant (Hubbard, Oertel, and Nummedal, 1979; Finley, 1978), and therefore tidal deltas in microtidal coasts tend to be flood oriented and build into adjoining lagoons. Most sand is concentrated on the barrier island itself, in flood-dominant tidal deltas, and in washover fans. Marshes tend to build on the lagoonal side of the barrier, whereas mud accumulates in the lagoon. The barrier islands of the coast of Texas, such as Galveston Island, are typical of barrier systems forming in microtidal coasts.

Mesotidal Coasts. Along mesotidal coasts, the wave regime may reach the same energy level as on microtidal coasts, but the overall tidal current velocity increases and tidal range increases from 2 to 4 m. Consequently, the role of tidal exchange becomes more significant. Morphologically, barrier islands on mesotidal coasts are characterized by short lengths (Figure 4.9); they tend to be wider on one end and narrow at the opposite end ("drumstick" barriers of Hayes, 1975, 1979), and they are cut by many more tidal inlets through which tidal exchange occurs (Figure 4.9). Tidal inlets in mesotidal coasts tend to be ebb dominant (Fitzgerald and Nummedal, 1983), and therefore tidal deltas are ebb dominant also and build on the ocean-side of barrier islands (Figure 4.9) by means of sediment bypassing (Fitzgerald, 1982). The growth of such ebb-dominated tidal deltas tends to deflect wave fronts, causing wave refraction around the tidal deltas. Consequently, localized longshore currents (Figure 4.10) are set up, producing a flow reversal opposite to main longshore current trends and aiding in depositing a larger volume of sediment on the so-called down side of the tidal delta. This deposition is aided by sediment by-

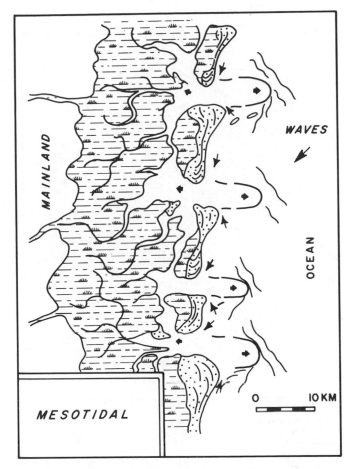

FIGURE 4.9. *Morphology of a typical mesotidal barrier island system showing larger number of tidal inlets and drumstick shape of barriers (redrawn from Hayes, 1979).*

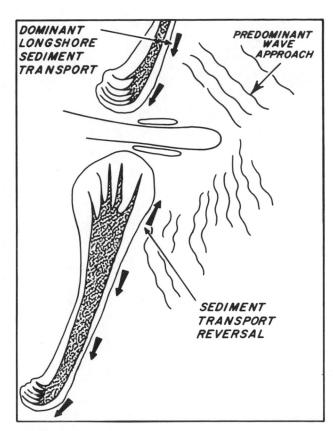

FIGURE 4.10. *Model showing orientation of predominant wave approach and resulting longshore current systems on mesotidal coast and correlation of changing flow pattern of longshore currents to development of drumstick shape of mesotidal barrier islands (redrawn from Hayes, 1979).*

passing through the tidal inlet (Fitzgerald, 1982); it is this additional sediment that causes the wider shape in the form of a drumstick. As one moves along the barrier island in the direction of wave approach, longshore currents are shore-parallel again and the barrier grows by progressive accretion by welding sediment onto existing berms (Hine, 1979), creating a beach-ridge morphology (Figure 4.10). The foreshore beach zone of mesotidal coasts is characterized also by a ridge-and-runnel morphology with a more complex internal structure.

Sand tends to accumulate on the drumstick-shaped barrier islands and in the ebb-dominant tidal deltas. Mud accumulates on the back-barrier side as tidal flats and marshes accrete and build over the high-tidal flat zones. The entire lagoon shore is fringed with tidal flats where mud and marsh sediment accumulate. The coast of South Carolina, USA, is a good example of a mesotidal coast (Hayes, 1979).

Macrotidal Coasts. Macrotidal coasts are tide dominated and wave subordinated, with tidal ranges exceeding 4 m. The coastal morphology consists of large and wide tidal flats. In some instance, such as in the Yellow Sea of Korea,

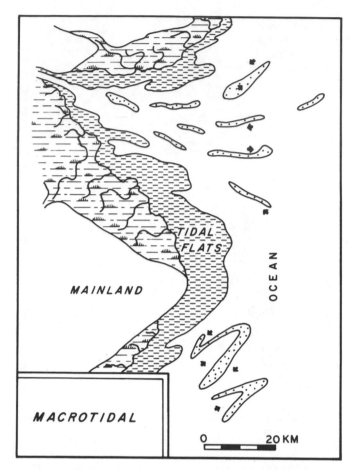

FIGURE 4.11. *Morphology of typical macrotidal coastline showing dominance of broad tidal flats, intertidal- and subtidal-current sand ridges and absence of barrier islands (redrawn from Hayes, 1979).*

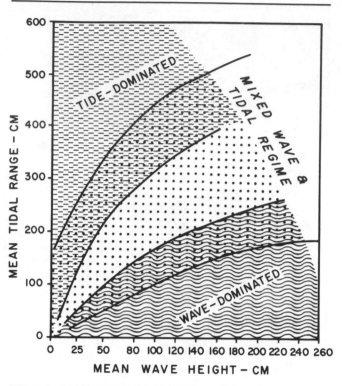

FIGURE 4.12. *Diagram showing relation of mean tidal range and mean wave height along a coastline and redefinition of different coastal process regimes. Coastal regimes are based on empirical observations from Bay of Fundy, Bristol Bay, Alaska, southwest Florida, northwest Florida, The German Bight, Copper River Delta, Alaska, Plum Island, Massachusetts, The Outer Banks of South Carolina, and Southeast Iceland (redrawn and modified from Hayes, 1979).*

tidal-flat widths reach from 8 to 25 km (Wells and Huh, 1980). As shown in the section on Intertidal Flats, the intertidal zone is subdivided according to changing velocity spectra during inundation, sediment distribution, and exposure (see Swinbanks, 1982). Thus, marshes tend to accumulate at and above high tide, mud accumulates on the higher elevations of a tidal flat, and sand on the lowest elevations (Figure 4.11). The bulk volume of sand does not accumulate on shore itself; instead it accumulates offshore as tidal-current sand ridges (Figure 4.11).

Features of macrotidal coasts are discussed in the sections on Intertidal Flats and Intertidal Sand Bodies. The coasts of the Yellow Sea, Korea, and the Bay of Fundy, Canada, (Chough, 1983; Klein, 1970a; Klein et al., 1982) are macrotidal.

Relative Role of Waves and Tides. The disadvantage of Davies's (1964) coastal classification is that it is based only on one parameter: tidal range. Observations (Figure 4.12) have demonstrated that both wave-energy flux, tidal range, and tidal-energy flux change independently along a coastline (Hayes, 1979). Thus, although one can categorize coasts as wave dominated and tide dominated (Figure 4.12), many coasts are characterized by seasonally changing mixtures of wave- and tidal-energy flux. Thus in some macrotidal coasts, winter storms would cause an increase in wave-energy flux and the spectrum of sediment distribution would change. Hayes (1969) demonstrated from a series of profiles obtained at Ipswich Beach, Massachusetts, that from late May through the summer season, these profiles changed from a microtidal slope to mesotidal ridges and runnels as the regime changed from wave dominated and tide subordinate to one that was tide dominant and wave subordinate. Clearly, this change came about because the relative wave energy diminished as tidal-energy flux became more effective. As shown in Figure 4.12, which is based on many sample points reported by Hayes (1979), the boundary conditions tend to be less distinct between different coastal dynamic modes than suggested by a classification based solely on tidal range. Even in mesotidal coasts where tidal-energy flux is large, flood-dominant tidal-inlet flow can cause flood-dominant tidal deltas to develop (Fitzgerald, Fink, and Lincoln, 1984). Thus caution in evaluating these models is advised because local and regional variation in processes changes sedimentation patterns.

BARRIER ISLAND TRANSGRESSIVE SEQUENCES

Our understanding of transgressive barrier island systems has come from detailed work along the Delaware Coast by Kraft (1971, 1978). Here, transgression and coastal erosion is displacing barrier systems landward by an overwash process (Kraft and John, 1979). However, coastal retreat along the Delaware coast shows little preservation of the barrier island by transgression. The effect of rising sea level in this case is to displace sand landward as overwash deposits over lagoonal sediments and to leave a thin sand veneer (approximately 30 cm thick) on the seaward side over the earlier positions of barriers. This sand veneer is the well-defined "ravinement" zone of Swift (1968) and it tends to be very thin when preserved at all. Kraft (1971, 1978) and Swift (1968) demonstrated that transgressive barrier systems show gener-

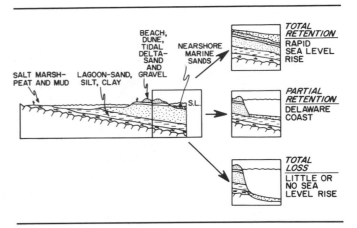

FIGURE 4.13. *Preservation potential for beach-barrier transgressive sediments as controlled by rate of relative sea level rise (redrawn from Kraft, 1971).*

ally a very small preservation potential. On the other hand, in a study of Gulf Coast Barriers along the Texas coast, Wilkinson (1975) demonstrated that a larger preservation potential of transgressive deposits exists prior to later seaward regression. Preservation would be enhanced by overwash processes of sedimentation (Kraft and John, 1979).

The variables that appear to control the preservation potential of transgressive coastal barriers are the rate of sea level rise and the rate of sediment yield (Kraft, 1971, 1978). In situations where the rate of sea level rise is large and rates of sediment accumulation are large, a complete transgressive sequence would be preserved (see Figure 4.13). Such a model was suggested by Fischer (1961) for the New Jersey coast. A partial sequence will be preserved where the rate of sea level rise is moderate and the rate of both sediment yield and accumulation is small. This combination of processes characterizes the present-day coast of Delaware (Figure 4.13; Kraft, 1971, 1978). Finally, in situations where the rate of sea level rise is small and the rate of sediment accumulation is also small, only a thin ravinement is left as a record of transgression (see Figure 4.13). A growing body of data suggests that this last circumstance, namely low rates of sea level rise and low rates of sediment accumulation, occurred more

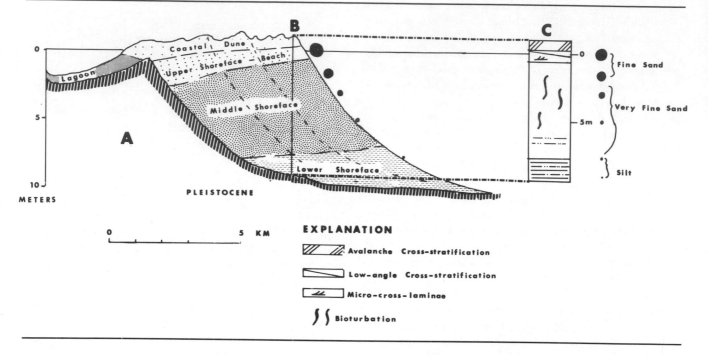

FIGURE 4.14. *Vertical and lateral sediment distribution of prograding barrier island based on Galveston Island model of Bernard, LeBlanc, and Major (1962). (A) Profile showing vertical and lateral sediment distribution and subenvironments. (B) Vertical section through prograded barrier island shown in C. (C) Vertical sequence of textures and sedimentary structures at B in prograding barrier island. (From Klein, 1974; republished with permission of the Geological Society of America.)*

frequently during the geologic past than previously proposed (see Ryer, 1977).

BARRIER ISLAND REGRESSION AND VERTICAL SEQUENCE

If the rise of sea level is low and sediment accumulation rates are large, or if sea level is stationary and sediment accumulation rates are large, or if sea level drops and sediment accumulation rates are moderate to large, coastal barrier islands will regress seaward toward the basin center. Nearly all preserved barrier systems in the rock record owe their origin to progradation as a consequence of one of these three combinations of depositional events. The best reference standards for prograding barrier island systems occur along the Gulf Coastal Plain of Texas (Bernard, LeBlanc, and Major, 1962). Their development since the major stillstand of sea level rise of about 3,500 to 5,000 BP has been documented by Bernard, LeBlanc, and Major (1962) and the associated sediment distribution is summarized in Figure 4.14. Progradation of such sequences generates a coarsening-upward sequence (Figure 4.14) that can be used to estimate water

depth (Klein, 1974). The coarsening-upward sequence consists of a basal offshore marine clay, an inner shoreface zone of interbedded muds and rippled and cross-stratified sands, an upper shoreface zone of beach sands with low-angle, accretionary cross-stratification, capped by eolian coastal sands with avalanche cross-stratification, and rootlet burrows at the top. These may be overlain by back-barrier lagoon or bay silts and clays. A summary sequence is shown in Figure 4.15.

Within the beach-dune complex itself, variations on this theme of coarsening-upward sequences may be observed. Thus, Barwis (1978) has reported that on Kiawah Island, South Carolina, a barrier island along a mesotidal coast, the overall coarsening-upward motif prevails, but

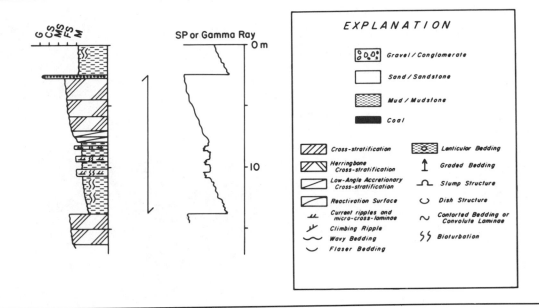

FIGURE 4.15. *Idealized vertical sequence and electric log shape of prograding microtidal coastal barrier island system. Doubly terminated arrow shows a complete interval. (Abbreviations: G—Gravel; CS—Coarse sand; MS—Medium sand; FS—Fine sand; M—Mud.)*

the relationships between preserved beach and overlying dune sediments differ from the Galveston Island model. Thus, on the landward side of Kiawah Island, marsh deposits are overlain by washover sediments from storms, wind flat sands, and berm crest and then dune deposits. On the seaward side, the middle shoreface sediments may grade upward into marsh deposits, the foreshore beach face with accretionary cross-stratification, runnel sediments with landward dipping avalanche cross-stratification, the berm crest, and overlying dunes with avalanche cross-stratification (Figure 4.16).

Figure 4.17 shows an ancient example of a regressive barrier island coarsening-upward sequence from the Atoka Formation (Pennsylvanian). A sedimentary log of several repeated coarsening-upward sequences from the Atoka Formation is shown in Figure 4.18; each sequence is separated from the next by a very thin (10 cm thick) ravinement zone representing transgressive events that occurred between each cycle of barrier regression.

ANCIENT COUNTERPARTS OF BARRIER ISLANDS

Barrier island sequences have been recognized in several ancient sedimentary rocks, including the Cretaceous of New Mexico (Campbell, 1971), the Cretaceous of Wyoming, and the Mississippian and the Silurian (Figure 4.19) of the Appalachians. The Mississippian and Pennsylvanian of the Black Warrior basin also contain barrier island sediments (Hobday, 1974).

Perhaps one of the better-studied examples of an ancient barrier island succession is Ryer's (1977) work on the Frontier Formation (Cretaceous) of northeastern Utah. There, coarsening-upward sequences representing prograding barrier island systems were observed in repeated sequences. Each is separated by an extremely thin transgressive ravinement sand. Figure 4.20 shows the mechanism for continued repetition in such asymmetrical regressive cycles. Preservation of the regressive sequences is favored over transgressive deposition. This field situation appears to be identical to the regressive barriers described from the Atoka Formation by Klein (1974) and shown in Figures 4.17 and 4.18.

These examples of asymmetric regressive cycles indicate that the preservation potential of coastal barrier systems is favored during regressive barrier sedimentation,

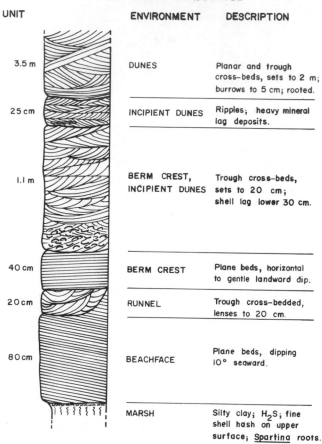

BEACH RIDGE LITHOLOGIES
SEAWARD SEQUENCE

UNIT	ENVIRONMENT	DESCRIPTION
3.5 m	DUNES	Planar and trough cross–beds, sets to 2 m; burrows to 5 cm; rooted.
25 cm	INCIPIENT DUNES	Ripples; heavy mineral lag deposits.
1.1 m	BERM CREST, INCIPIENT DUNES	Trough cross–beds, sets to 20 cm; shell lag lower 30 cm.
40 cm	BERM CREST	Plane beds, horizontal to gentle landward dip.
20 cm	RUNNEL	Trough cross–bedded, lenses to 20 cm.
80 cm	BEACHFACE	Plane beds, dipping 10° seaward.
	MARSH	Silty clay; H_2S; fine shell hash on upper surface; Spartina roots.

FIGURE 4.16. *Vertical sequence on seaward side of a barrier island complex, Kiawah Island, South Carolina (from Barwis, 1978; republished with permission of Southeastern Geology).*

FIGURE 4.17. *Coarsening-upward sequence produced by prograding barrier island, Atoka Formation (Pennsylvanian), Winn Mountain, Arkansas. (1) Coastal dune sandstone. (2) Beach-shoreface sandstone. (3) Middle shoreface. (4) Lower shoreface mudstone. (From Klein, 1974; republished with permission of the Geological Society of America.)*

whereas it is extremely small during transgressive coastal sedimentation. Presumably the reason for the small preservation potential of transgressive barriers is the combination of a low to moderate rate of sea level rise with small rates of sediment accumulation. In the Atoka Formation, it is known that deltaic deposition occurred simultaneous with beach-barrier deposition (Klein, 1974). Therefore, one may assume that the deltas acted as sources for sand that was transported by longshore currents to form the barriers. Imbrication of the deltas, which occurs today in the Mississippi Delta (Coleman, 1976; Scruton, 1960), would account, potentially, for periodic elimination of sources of sediment yield to the barriers. If such imbrication is coupled with a slow rate of sea level rise, it would be expected, based on Kraft's (1971, 1978) analysis, that only ravinement sands would be preserved as an episode of transgression. It is noteworthy, as Ryer (1977) showed, that perhaps the transgressive event may well have lasted one or two orders of magnitude longer than the time needed to accumulate the regressive barrier cycle.

MACROTIDAL SAND BODIES

This section reviews the salient features of sandy tidalites, which are sandy sediments deposited by tidal currents (Klein, 1971) and accumulated on macrotidal coasts (Figure 4.11). Such sands occur in regions where tidal-current systems are dominant. Tidal-current systems are known to occur as depositional agents in a variety of water depths ranging from tidal flats at sea level to depths of 2,000 and 2,500 m (Keller et al., 1973; Shepard and Marshall, 1973; Lonsdale, Normark, and Newman, 1972; Lonsdale and Malfait, 1974; Shepard et al., 1979). Thus, the deposition of sandy tidalites can occur in a variety of macrotidal coastal settings including deltas, barrier island complexes (especially tidal inlets), tidal marshes, and tidal deltas. The dominant coastal setting where such sediments occur is the tidal flats.

On continental shelves, tidal currents are a common process, and both tidal-current systems and tidal sand-body accumulation become dominant where wide continental shelves occur (Redfield, 1959; Off, 1963). As documented many years ago by Sverdrup, Johnson, and Fleming (1942) and later by Redfield (1958) and more recently by Klein (1977a,b), Klein and Ryer (1978), and Cram (1979), there is a direct positive correlation between increasing shelf width and increasing tidal range (see also Figure 6.3 and p. 123). As a consequence, on wide continental shelves tidal-current systems are dominant and extensive, and large subtidal, tide-dominated sand bodies (or tidal-current sand ridges) accumulate. Most macrotidal coasts are associated with wide continental shelves and embayed coasts on narrower shelves (Hayes, 1979).

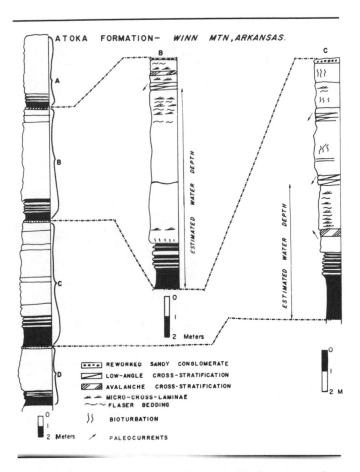

FIGURE 4.18. *Sedimentary log showing vertical sequence of lithologies and sedimentary structures in four succeeding prograding barrier island sequences, Atoka Formation (Pennsylvanian), Winn Mountain, Arkansas. Interval for estimating water depth as given in Klein (1974) shown in expanded logs for (B) and (C). Thin ravinement zones indicated as reworked sandy conglomerate. Black = mudstone; white = sandstone. (From Klein, 1974; republished with permission of the Geological Society of America.)*

It must be stressed again that many geologists assume that tidal-sediment processes and deposition deal only with tidal-flat settings (see discussions by Klein, 1976, p. 2; 1977a, p. 2). This assumption overlooks the significance of tidal-sediment transport and depositional processes occurring over very large areas of the world such as the broad continental shelves, and perhaps their ancient cratonic platform counterparts (Bowsher, 1967;

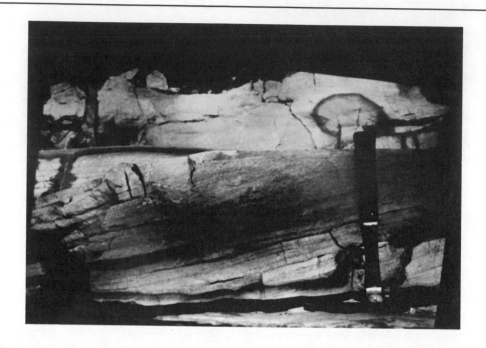

FIGURE 4.19. *Low-angle accretionary cross-stratification in Clinton Group (Silurian), Frenchman Mountain, Pennsylvania, showing preservation of beach foreshore facies.*

Klein, 1977b; Klein and Ryer, 1978). Such subtidal areas are in fact tide dominated and the areal extent of the sediment accumulations, particularly the tidal sand bodies, is extensive. The preservation potential of such tidal sand bodies in the stratigraphic record is therefore considerably enhanced and it is fair to reiterate (Klein, 1977a) that these sediment types are more common in the rock record than previously supposed. As a result, one should expect that a larger number of them comprise major reservoirs of oil and gas (see also W.E. Evans, 1970; Berg, 1975; Brenner, 1978).

INTERTIDAL FLATS

The intertidal-flat zone occurs on coastlines between high and low tide level. It is a broad plane of deposition occurring along mesotidal and macrotidal coastlines of low relief as well as along rocky coasts.

Studies of clastic tidal flats go fairly far back; these were reviewed by Van Straaten (1961), Reineck (1967), Hantzschel (1939), and Klein (1976, 1977a, 1984). The type areas for clastic intertidal-flat studies include the interti-

dal flats of the North Sea coast of the Netherlands (Van Straaten, 1952, 1954, 1959, 1961), of northwest Germany (Hantzschel, 1939; Reineck, 1963, 1967, 1972), and of The Wash in eastern England (G. Evans, 1965). There, the overall sediment textural distribution coarsens seaward from high water to low water (see summary in Klein, 1971, 1972a,b, 1977a, 1985a) and three tidal-flat zones are recognized. These zones are (from high tide to low tide) the high tidal flat, the mid-tidal flat and the low tidal flat. Landward of the high tidal flat are supratidal salt marshes, and seaward of the low tidal flat is the shallow, tide-dominated subtidal zone. Both the tidal marshes and the intertidal flats are incised by another depositional zone, the tidal channels (Figure 4.21).

Coarsening-seaward textural distributions have been documented from tidal flats in other parts of the world, including the Gulf of California (R.W. Thompson, 1968), the Minas Basin of the Bay of Fundy (Klein, 1963, 1970a),

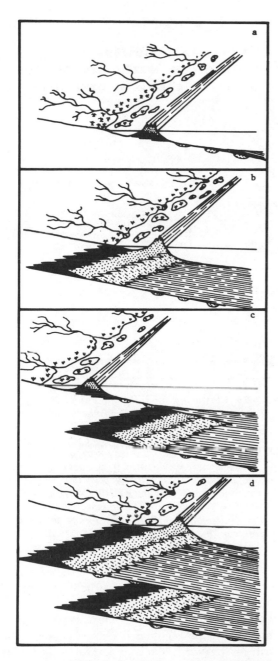

FIGURE 4.20. *Evolution of multiple coarsening-upward sequences by barrier island regression (from Ryer, 1977; republished with permission of the Geological Society of America).*

San Francisco Bay (Pestrong, 1972), the coast of British Columbia (Kellerhals and Murray, 1969; Swinbanks and Murray, 1981), the coast of Massachusetts (Hayes, 1969), Inhaca Island, Mozambique (Hobday, 1977), and North Australia (Semeniuk, 1981). Local changes will, however, provide minor exceptions to this motif of sedimentation, such as in the Bay of Fundy where sea cliffs of bedrock occur and a fringing gravel zone forms at high tide (Klein, 1963, 1970a) or in the northwest Gulf of California where alluvial fan sediments intertongue with tidal-flat sediments, some of which contain evaporite minerals (R.W. Thompson, 1968). Overall, the seaward-coarsening textural motif characterizes clastic tidal flats in a variety of settings and it is the definitive feature of this depositional environment.

Besides this overall textural trend, the association of grain-size distributions and sedimentary structure types of each of the tidal-flat subzones is also distinctive. These are reviewed below:

High Tidal Flat. The high tidal flat is dominated by mud deposition with minor amounts of parallel laminae and bioturbation. Exposure features, including mudcracks and wrinkle marks, are common. This zone of the tidal flat is inundated for the shortest period of a tidal cycle coincident with slack water velocities at high tide. Only suspension depositional processes occur under these conditions and therefore, the high tidal flat environment is a suspension-dominated environment (Figure 4.21).

Mid-Tidal Flat. The mid-tidal flat environment consists of coarser sediment of nearly equal proportions of mud and sand arranged into lenticular, flaser, tidal, and wavy bedding. Current and interference ripples and exposure features, such as mudcracks and wrinkle marks, are common. This portion of the tidal flat is inundated for an average of half the total duration of the tidal cycle and receives a nearly equal amount of both bedload and suspension deposition, which results in the mixed lithologies and associated features listed above. In terms of sediment transport, the mid-tidal-flat environment is a mixed environment, transitional in terms of bedload and suspension deposition (Figure 4.21).

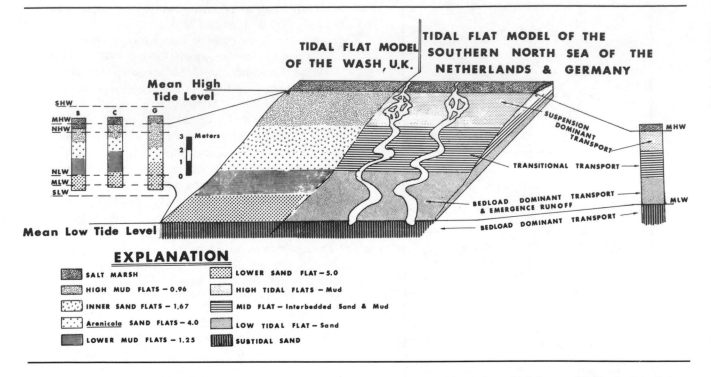

EXPLANATION

SALT MARSH	LOWER SAND FLAT — 5.0
HIGH MUD FLATS — 0.96	HIGH TIDAL FLATS — Mud
INNER SAND FLATS — 1.67	MID FLAT — Interbedded Sand & Mud
Arenicola SAND FLATS — 4.0	LOW TIDAL FLAT — Sand
LOWER MUD FLATS — 1.25	SUBTIDAL SAND

Low Tidal Flat. The low tidal-flat environment consists dominantly of sand-sized sediment, fashioned into dunes and sand waves with internal herringbone or unimodally oriented cross-strata with reactivation surfaces. Current ripples mantle the surface of these bedforms. Exposure features, such as mudcracks and wrinkle marks (Runzel marks), occur in this setting, but their preservation potential is low. A variety of emergence runoff features, such as washout and current ripples superimposed on dunes and sand waves, are common to this setting. This environment is inundated for the longest period of the tidal cycle and is subjected to the greatest bottom-tidal-current velocities. Consequently, the dominant mode of sediment transport and deposition is bedload sedimentation and emergence runoff (Figure 4.21).

The coarsening-seaward textural distribution, so diagnostic of this environment, owes its origin, then, to the combination of differential times of inundation and submergence of tidal flats during a tidal cycle, and the associated change of the bottom-current velocity spectrum during the same cycle. A more detailed discussion with

FIGURE 4.21. *Clastic intertidal-flat models for the North Sea Coast of the Wash (Evans, 1965), the Netherlands (Van Straaten, 1954), and Germany (Reineck, 1963, 1967) showing sediment distribution, sediment transport zones, vertical sequences, sand-mud ratios (numbers were pertinent), and fining-upward sequences generated by prograding. Actual core logs shown for Wash. Abbreviations: SHW (Spring High Water); MHW (Mean High Water); NHW (Neap High Water); NLW (Neap Low Water); MLW (Mean Low Water); SLW (Spring Low Water). (From Klein, 1971; republished with permission from the Geological Society of America.)*

illustrations of the sedimentary structures appears in Klein (1977a, 1985a), to which the reader is referred.

It should be stressed in this context, however, that the generalized scheme mentioned above, which is based primarily on the North Sea, is in part an artifact of availability of sediment ranging from clay to sand. There are cases known (Wells, 1983; Wells, Prior, and Coleman, 1980) where extensive muddy intertidal flats occur, some containing extensively fluidized muds that move as a mobile mass. These muddy sediments attenuate incoming waves and thus dampen the orbital motion, minimizing

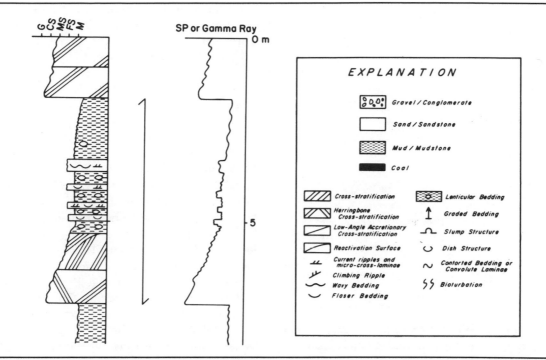

FIGURE 4.22. *Idealized vertical sequence and predicted log shape of prograding intertidal flat. (Abbreviations: G—Gravel; CS—Coarse sand; MS—Medium sand; FS—Fine sand; M—Mud.)*

erodibility of sediment. Thus progradation rates tend to become rapid. Localized slumping on extremely shallow slopes is known to occur (Wells, Prior, and Coleman, 1980). A more general zonation scheme was proposed recently by Swinbanks (1982) based on longer-term exposure of individual parts of the intertidal zone. This zonation is based on so-called Critical Tide Levels. The amphizone, which is in the middle of the intertidal zone, experiences changes in submergence and exposure in hours, whereas the overlying atmozone and underlying aquazone experience exposure and submergence extremes that are on the order of magnitude of days. This zonation fits much more closely to the biological changes on an intertidal flat.

TIDAL-FLAT VERTICAL SEQUENCE

The Holocene tidal flats of the North Sea of western Europe are characterized by a history of progradation. In the Netherlands, the Holocene stratigraphic record shows at least three such progradational cycles (DeJong, 1965; Hageman, 1972). When progradation occurs, each of the subenvironments of the tidal flats accumulates sediment in a seaward direction and oversteps the seawardmost environment adjacent to it. Thus, high tidal-flat muds will prograde over the interbedded sands and muds of the mid-tidal flat. Mid-flat sediments, in turn, prograde over low tidal-flat sands. Continued progradation generates a vertical sequence (Klein, 1971, 1972a,b, 1977a), which fines upward (Figures 4.21 and 4.22). The vertical sequence, from the base upward, consists of lower tidal-flat sands, mid-flat interbedded sands and muds, and high tidal-flat muds. These sequences may be overlain by thinly preserved supratidal marsh deposits or supratidal muds with ironstone concretions (Eriksson, 1977; Tankard and Hobday, 1977). The thickness of Holocene fining-upward sequences coincides with Holocene tidal range (Klein, 1971), inasmuch as compaction is an early diagenetic event that precedes progradation (Klein, 1972a) and is therefore not expected to modify the preserved thickness of the sequence. In ancient counterparts,

TABLE 4.3. *Rates of known progradation of Holocene intertidal flats*

Location	Rate	Reference
Netherlands		Hageman, 1972
G-4 Interval	5 m/yr	
G-3 Interval	8 m/yr	
G-2 Interval	4 m/yr	
G-1 Interval	2.7 m/yr	
Average for Netherlands	4.9 m/yr	
France	1 m/yr	LeFournier and Friedman, 1974
Alaska	12 m/yr	Ovenshine et al., 1975; Ovenshine, Lawson, and Bartsch-Winkler, 1976

the thickness of such sequences can approximate paleo-tidal range. Lateral variability in ancient counterparts is apparently minimal (Klein, 1972b) because tidal flats along open coasts appear to show the best preservation potential; such coasts would be relatively straight, thus minimizing the lateral variation of tidal range so characteristic of indented coasts.

Evidence exists for rapid rates of tidal-flat progradation in two areas. LeFournier and Friedman (1974) recorded an average rate of tidal-flat progradation of 1 km per century (1 m/yr) in northern France. Ovenshine, Lawson, and Bartsch-Winkler (1976) and Ovenshine et al. (1975) reported that during the 1964 Alaska Earthquake, the intertidal-flat environment of Turnagain Arm, Alaska, subsided approximately 1.5 to 2.0 m. Since 1964, an additional 2.0-m-thick sequence of fining-upward tidal-flat sediments has prograded over a depositional zone that is 1.8 km wide, representing a calculated progradation rate of 12 m/yr. This post-1964 sequence overlies a slightly older fining-upward tidal-flat sequence without a ravinement. These data on progradation rates, plus the known repetition and preservation of four Holocene progradational sequences of tidal flats in the Netherlands, which yield an average calculated rate of progradation of 4.9 m/yr (DeJong, 1965; Hageman, 1972), indicate that in areas of large sediment accumulation rates, such fining-upward sequences would be preserved. Table 4.3 summarizes these progradation rates.

INTERTIDAL SAND BODIES

Along macrotidal coasts, large intertidal sand bodies have been reported, particularly from the Minas Basin of the Bay of Fundy (Klein, 1970a; Knight and Dalrymple, 1975; Dalrymple, Knight, and Lambiase, 1978; Lambiase, 1980), King Sound, northern Australia (Gellatly, 1970; Semeniuk, 1981), and Garolim Bay, Korea (Song, Yoo, and Dyer, 1983). These sand bodies tend to accumulate along broad tidal flats and are, in essence, a counterpart to the low tidal-flat environment. However, their morphology and history differ in detail. Their size and potentially favorable porosity and permeability, which make them of interest as potential reservoirs of oil and gas, provide the basis for including them herein as a separate group.

The Minas Basin of the Bay of Fundy is characterized by the highest tidal range in the world. Tidal currents over the intertidal sand bodies are characterized by the property of time–velocity asymmetry; that is, velocities are greater during one phase of the tidal cycle (dominant phase) than the alternating (subordinate) phase (Figure 4.23). In the Minas Basin, bottom-current velocities reach a maximum of 120 cm/sec. Tidal-current systems are thoroughly mixed with ebb-current systems flowing generally westward and flood-current systems flowing easterly (Figure 4.24).

Sediment distribution generally coarsens seaward, except for gravel along bedrock ledges and sea cliffs (Figure 4.25). The bar morphologies are nearly identical: asymmetrical in cross-section and linear in plan (Figure 4.26). The distribution of time–velocity asymmetry zones coincides with both bar topography (Figures 4.26 and 4.27), the distribution of sedimentary facies (Figure 4.28), and the orientation of dune and sand-wave bedforms (Figure 4.29). Although grain dispersal from point sources is known to be both radial and elliptical, the direction of maximum and mean distances of grain dispersal also coincides with the distribution of time–velocity asymmetry zones. As a consequence, grain dispersal is through al-

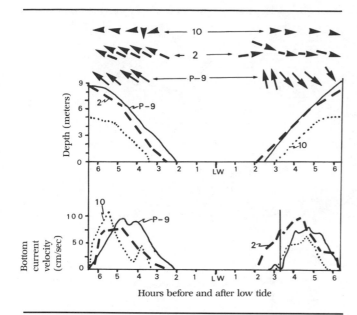

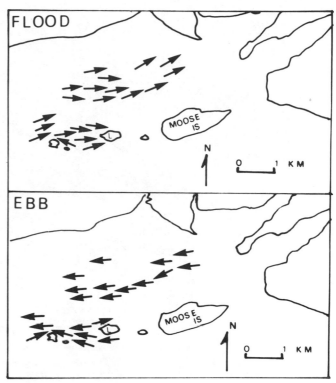

FIGURE 4.23. *Record of 13-hr continuous observations of bottom-current velocities at flood-dominated steep face of tidal sand body (Station #2), and ebb-dominated gently sloping surfaces of tidal sand body (Stations #10 and P-9), West Bar, Economy Point, Minas Basin, Bay of Fundy, Nova Scotia (from Klein, 1970a; republished with permission of the Society of Economic Paleontologists and Mineralogists).*

FIGURE 4.24. *Orientation of flow directions of bottom-tidal currents at Five Islands, Nova Scotia (from Klein, 1970a; republished with permission of the Society of Economic Paleontologists and Mineralogists).*

ternate ebb-dominated and flood-dominated time–velocity asymmetry zones (Figure 4.30). Time–velocity symmetry also produces reactivation surfaces (Figure 4.31) as part of the internal anatomy of dunes and sand waves. This dispersal mode controls sand-body alignment, which is parallel to both the direction of tidal-current flow and depositional strike.

Sedimentary structures present on intertidal sand bodies include dunes and sand waves with internal cross-stratification, reactivation surfaces, various types of current ripples, including double-crested current ripples, and oblique orientations of current ripples superimposed on dunes and sand waves. Internally "B-C" sequences (Klein, 1970b) of micro-cross-laminae overlying cross-strata and reactivation surfaces are present. The orienta-

tion of dunes, sand waves, and cross-stratification is also aligned parallel to depositional strike and sand-body alignment. A more detailed treatment of the relationship of sedimentary structure formation and specific modes of tidal-current deposition is provided in Klein (1971, 1977a, 1985a).

More recent work (Visser, 1980; Boersma and Terwindt, 1981) demonstrated that during a tidal cycle, longer-term changes in the magnitude of bottom-current velocities are observed during the shift from neap to spring tide and back to neap tide. Under such changes, distance of bedform migration increases during spring tides and is either minimal or approaches zero during neap tides. Thus, larger-order bounding surfaces are seen

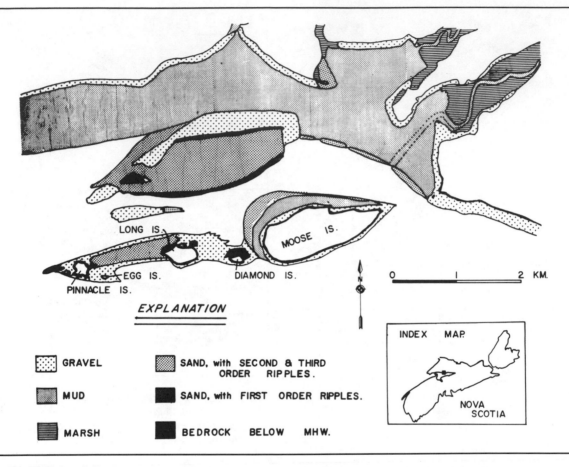

LONG IS.

MOOSE IS.

EGG IS.   DIAMOND IS.

PINNACLE IS.

N

0   1   2 KM.

EXPLANATION

INDEX MAP.

NOVA SCOTIA

:::: GRAVEL

▨ MUD

≡ MARSH

▦ SAND, with SECOND & THIRD ORDER RIPPLES.

■ SAND, with FIRST ORDER RIPPLES.

■ BEDROCK BELOW MHW.

FIGURE 4.25. *Sediment distribution map, Five Islands, Minas Basin, Nova Scotia (from Klein, 1970a; republished with permission of the Society of Economic Paleontologists and Mineralogists).*

to demarcate bundles of sets of cross-stratified sediments that represent bedform migration and accumulation during the dominant phase of tidal-current transport during spring tide. The pause-planes, or thin accumulations of sediment deposited during neap tide, offset these larger bundles of greater volumes of sand with cross-stratification on a lateral view. These features permit one to discriminate both neap and spring tide features respectively in both Holocene sediments and ancient counterparts (Boersma and Terwindt, 1981; Allen, 1982; see Klein, 1985a for a more detailed summary).

VERTICAL SEQUENCE OF INTERTIDAL SAND BODIES

The macrotidal coast of the Minas Basin of the Bay of Fundy appears to be one of equilibrium between sedi-

ment erosion and accumulation (Klein, 1970a; Knight and Dalrymple, 1975). Both Klein (1970a, 1972a, 1977a) and Knight and Dalrymple (1975) stressed that this coast is not progradational and thus, no vertical sequence can be observed directly by means of coring. A hypothetical facies sequence was proposed by Knight and Dalrymple (1975), however, consisting of a thick basal intertidal sand body, overlain by braid-bar sand and capped by silts and muds of supposed high tidal flats (Figure 4.32). Their fining-upward sequence differs from the prograding tidal-flat sequence discussed in the previous section by being dominated by a thick basal sand representing the

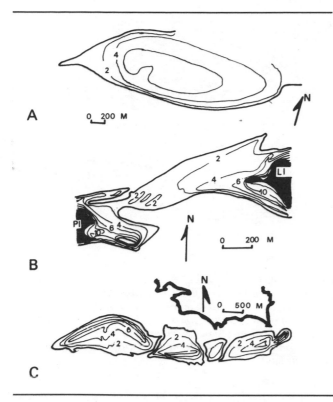

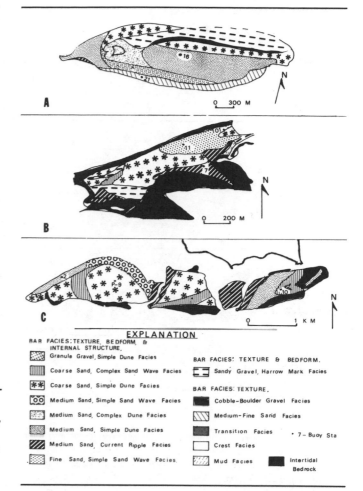

FIGURE 4.26. *Tidal sand-body topography for (A) Big Bar, Five Islands, (B) Pinnacle Flats, Five Islands, and (C) Economy Point, Minas Basin, Bay of Fundy. CI = 2 m. (From Klein, 1970a; republished with permission of the Society of Economic Paleontologists and Mineralogists.)*

FIGURE 4.27. *Areal distribution of time–velocity asymmetry zones for (A) Five Islands, and (B) Economy Point, Minas Basin, Bay of Fundy (from Klein, 1970a; republished with permission of the Society of Economic Paleontologists and Mineralogists).*

FIGURE 4.28. *Distribution of sedimentary facies at (A) Big Bar, (B) Pinnacle Flats, and (C) Economy Point, Minas Basin, Bay of Fundy, Nova Scotia (from Klein, 1970a; republished with permission of the Society of Economic Paleontologists and Mineralogists).*

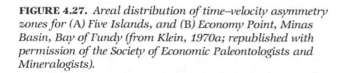

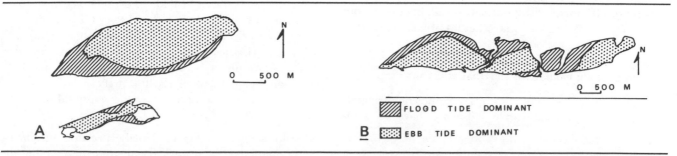

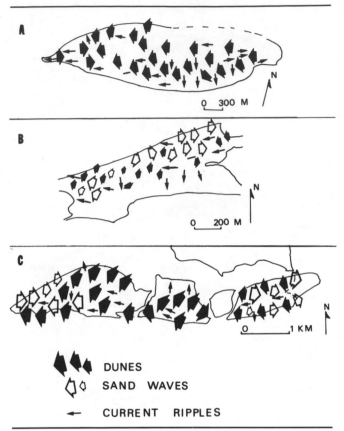

DUNES

SAND WAVES

← CURRENT RIPPLES

FIGURE 4.29. *Orientation of bedforms at (A) Big Bar, (B) Pinnacle Flats, and (C) Economy Point, Minas Basin, Bay of Fundy, Nova Scotia (from Klein, 1970a; republished with permission of the Society of Economic Paleontologists and Mineralogists).*

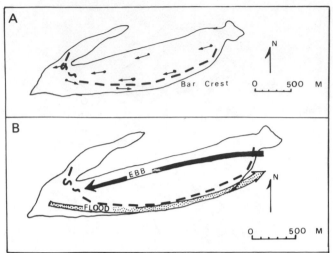

FIGURE 4.30. *(A) Direction of maximum dispersal of sand grains from point source after one tidal cycle at Big Bar, Five Islands, Nova Scotia. (B) Generalized sand dispersal circulation model through zones of flood- and ebb-dominated tidal currents (from Klein, 1970a; republished with permission of the Society of Economic Paleontologists and Mineralogists).*

intertidal sand body. It is similar, however, because it does fine upward. This hypothetical fining-upward sequence is unproven in the Holocene. An ancient example was reported, however, by Barnes and Klein (1975) from the Cambrian Zabriskie Quartzite of California and Nevada.

ANCIENT COUNTERPARTS OF INTERTIDAL FLATS AND SAND BODIES

Since the late 1960s, many examples of ancient intertidal-flat and intertidal sand-body sediments have been re-

ported. Klein (1970b, 1977a,b) and Ginsburg (1975) compiled many examples based on comparative descriptive sedimentology relying mainly on association of sedimentary structures. Klein and Ryer (1978) cited several examples from South Africa, including the Moodies Supergroup (Eriksson, 1979), the Pongola Group, and the Pretoria Group. In North America, the Middle Member of the Wood Canyon Formation (Precambrian), the Zabriskie Quartzite (Cambrian), the Eureka Quartzite (Ordovician), and the Cretaceous units in the Rocky Mountains such as the Almond Formation, Dakota Sandstone, Fox Hills Sandstone, and Viking Formation (Evans, 1970) were also mentioned by Klein and Ryer (1978). Other examples include the Cambrian Mt. Simon Formation of Wisconsin (Driese, Byers, and Dott, 1981) and the Precambrian Palms and Pokegama formations of the Lake Superior region (Ojakangas, 1982). Perhaps one of the best examples is the Lower Jurassic intertidal flat facies exposed at Bornholm, Denmark (Sellwood, 1972).

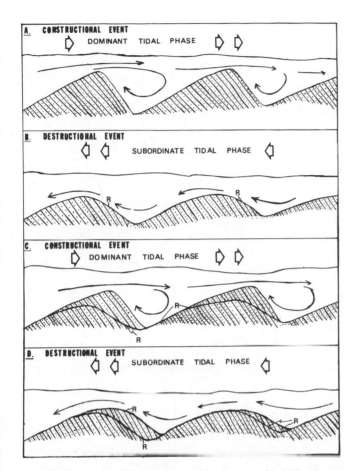

FIGURE 4.31. *Genetic model for development of reactivation surfaces during alternation of dominant tidal phase (construction phase) when dunes migrate, and subordinate phase (destructional event) when reactivation surfaces (R) develop. This model pertains only to areas of tidal currents characterized by time–velocity asymmetry. Repeated alternation produces multiple reactivation surfaces (from Klein, 1970a; republished with permission of the Society of Economic Paleontologists and Mineralogists).*

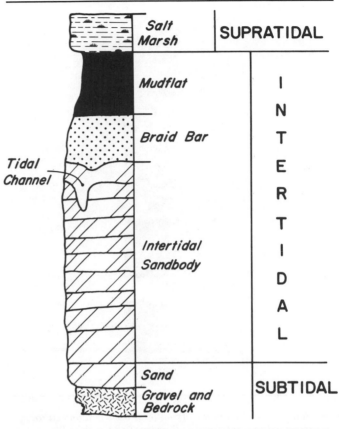

FIGURE 4.32. *Hypothetical vertical sequence of prograding macrotidal coast with intertidal bodies (redrawn from Knight and Dalrymple, 1975, and Klein, 1977a).*

OIL, GAS, AND URANIUM EXAMPLES

BARRIER ISLAND HYDROCARBON RESERVOIRS AND URANIUM DEPOSITS

Several excellent examples of barrier island reservoir sands have been documented. Among the classic examples are the Pennsylvanian "shoe-string" sands of Kansas, the Bisti Field of New Mexico, the Deadhorse Creek Field of Wyoming (Sabins, 1962), and the well-known Bell Creek Field of Wyoming (Berg and Davies, 1968; Davies, Ethridge, and Berg, 1971). More recently, the "J" Sandstone of the Third Creek Field of Colorado (Cretaceous)

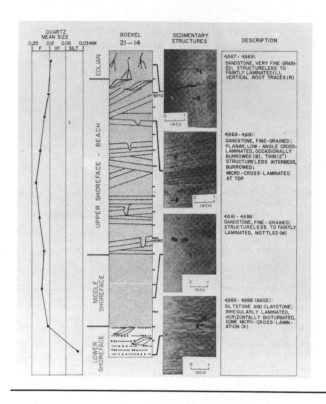

FIGURE 4.33. *Sedimentary structures, textures, and lithology of barrier island reservoir sediments, Bell Creek Field, Montana, Well 6-14 (from Davies, Ethridge, and Berg, 1971; republished with permission of the American Association of Petroleum Geologists).*

described by Reinert and Davies (1975) and the sands of the Cotton Valley Group (Cretaceous) of Louisiana (Sonnenberg, 1977) were shown to contain petroleum reservoirs in ancient barrier sequences.

The Bell Creek example is perhaps one of the better-documented case histories (Berg and Davies, 1968; Davies, Ethridge, and Berg, 1971). There coarsening-upward cycles with associated structures were observed in cores recovered from the field (Figure 4.33), and the electric log signature (Figure 4.34) is one of a transitional base and a blunt top (see also Selley, 1976, and Figure 1.1). This signature characterizes all the examples listed above.

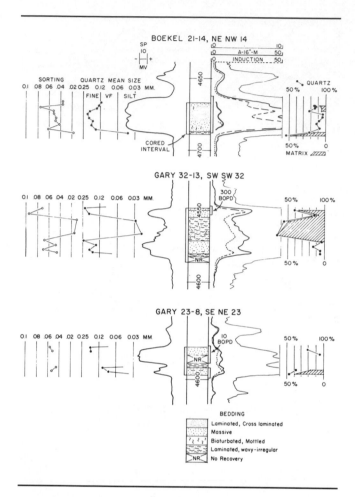

FIGURE 4.34. *Electric log characteristics of Muddy Sandstone Facies representing barrier islands, Bell Creek Field, Montana (from Berg and Davies, 1968; republished with permission of the American Association of Petroleum Geologists).*

The seismic signature of such barrier sequences is, again, a function of acoustic impedance. Figure 4.35 shows a synthetic seismic cross-section developed by Schramm, Dedman, and Lindsey (1977) for a coastal barrier complex.

There are several known occurrences of barrier sandstone-type uranium deposits also. The better ones have been described from the Gulf Coastal Plain of Texas from

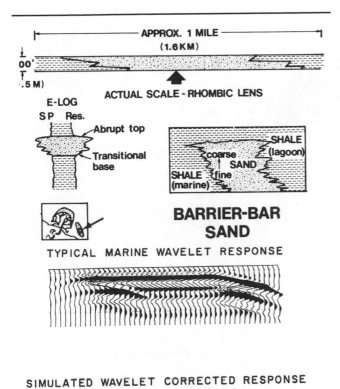

FIGURE 4.35. *Geological, electric log, typical marine wavelet response, and simulated wavelet corrected seismic response for barrier island sandstone system (from Schramm, Dedman, and Lindsey, 1977; republished with permission of the American Association of Petroleum Geologists).*

of uranium available. In this case, interbedded volcanic ash beds acted as a source, and as Galloway (1977) suggested for the fluvial examples discussed in Chapter 2, alkaline-oxidizing waters would leach that uranium and transport it through the barrier sands. Impedance of ground-water flow and the presence of the organic matter would then favor precipitation of the uranium on the updip side (Figure 4.36).

INTERTIDAL FLAT AND CHANNEL HYDROCARBON AND URANIUM OCCURRENCES

Almost no examples of intertidal-flat and intertidal-channel hydrocarbon reservoirs are known. Only one is noteworthy, the intertidal channel reservoirs associated with the Niger Delta (Weber, 1971, his Figure 13) from the Egwa Oil Field. This tidal channel also tongues laterally with tidal inlets that breach associated barrier islands, as discussed in Chapter 5. A second example that has been offered is some tidal-channel sandstone reservoirs associated with the Aux Vases Sandstone (Mississippian) of the Illinois Basin (Weimer, Howard, and Lindsay, 1982). However, more recent work by Seyler (1984) showed these sandstones to be counterparts to tidal-current sand ridges.

As mentioned in Chapter 2, uranium occurrences in sandstone appear to be confined to sandstones derived from a granitic or volcanic source containing uranium, and interbedded with mudstone. The uranium accumulation is diagenetic, controlled by alkaline-oxidizing ground-water circulation that transports uranium in solution through sandstones. Precipitation of uranium in sandstones is accelerated by impedance to ground-water flow at the sandstone-mudstone intertongues, or by interbedded organic-rich layers.

The interbedding of sandstones and mudstones is common to tidal flats, particularly where the intertidal-channel fill consists of sands. The supratidal environment contains salt marshes that serve, potentially, as an organic precipitant for uranium in solution. In short, the tidal-flat environment contains all physical attributes that are unique to sandstone-type uranium deposits. If such sandstones are derived from volcanic and granitic sources, or are interbedded with volcaniclastic sediments, uranium should expect to be deposited diagenetically in such sands.

the Jackson Group (Dickinson, 1976; Dickinson and Sullivan, 1976; Fisher et al., 1971). There again, uranium tends to be concentrated in sandstones where they are in contact with mudstones and organic-rich beds. The largest uranium concentrations tend to occur on the updip side of such barrier sands (Fisher et al., 1971), presumably because the updip sides of barrier systems tongue into lagoonal muds and marshes (Figure 4.36). This configuration would be favorable to uranium accumulation (see discussion in Chapter 2), provided there is a source

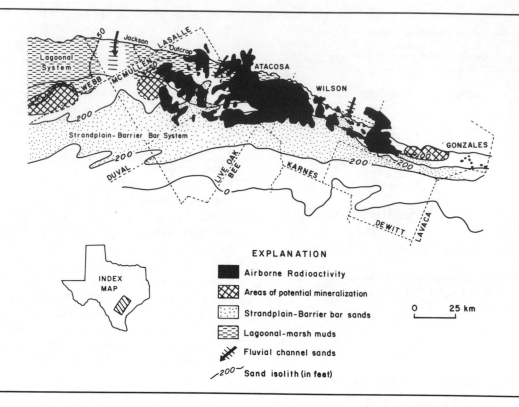

EXPLANATION

⬛ Airborne Radioactivity

▨ Areas of potential mineralization

⠿ Strandplain-Barrier bar sands

▤ Lagoonal-marsh muds

⬇ Fluvial channel sands

⌇200⌇ Sand isolith (in feet)

0   25 km

Three recent cases of such sandstone-type uranium deposits have been reported from intertidal-channel fills. The first example is from the Eocene Whitsett Formation of Texas (Dickinson and Sullivan, 1976), where tidal-channel fills contain uranium. The second example is from the Permian Cloud Chief Formation of Oklahoma (Al-Shaieb et al., 1977, p. 373), where uranium occurs in lenticular, porous siltstone considered to be tidal-channel fills. Similar uranium occurrences have been reported also by Al-Shaieb et al. (1977) from the Permian Wellington Formation of Oklahoma.

FIGURE 4.36. *Distribution of known and potential areas of uranium mineralization in Eocene Jackson Group of Gulf Coast Plain of Texas. Uranium concentration appears confined to updip side of strandplain complex (redrawn from Fisher et al., 1971).*

Chapter 5

Deltas

INTRODUCTION

The deltaic sand-body environment represents one of the most complex depositional systems reviewed in this monograph. The complexity of deltas owes its origin directly to the relative interplay of riverine discharge, wave-energy flux, tidal-energy flux, sediment dispersal, climate, and tectonic setting, among other variables (Galloway, 1975; Coleman, 1976, 1980; Coleman and Wright, 1975).

Deltas are fan-shaped sedimentary bodies in plan and lenticular in cross-section. They represent a form response to the sudden decrease in fluvial discharge and sediment discharge where a river system enters a standing body of oceanic or lake water. The sediments comprising deltaic systems owe their origins to a sudden decrease in hydraulic gradient from river influx and subsequent or advective reworking and redistribution of such sediment by wind-driven wave systems, tidal currents, or shelf currents.

Deltas have been classified according to the relative dominance of fluvial discharge, wave action, and tidal action (Figure 5.1). The delta of the Mississippi River represents an excellent example of a river-dominated delta, whereas the deltas of the São Francisco River of Brazil and of the Senegal River of Africa are excellent examples of wave-dominated deltas (Coleman, 1980). The Ganges-Brahmaputra Delta of Bangladesh, the Klang-Langat Delta of Malaysia, and the Fly River Delta of Papua represent examples of tide-dominated deltas. Intermediate deltas are also known (Figure 5.1) including the Danube (wave-river), Copper (wave-tide; Galloway, 1976), and

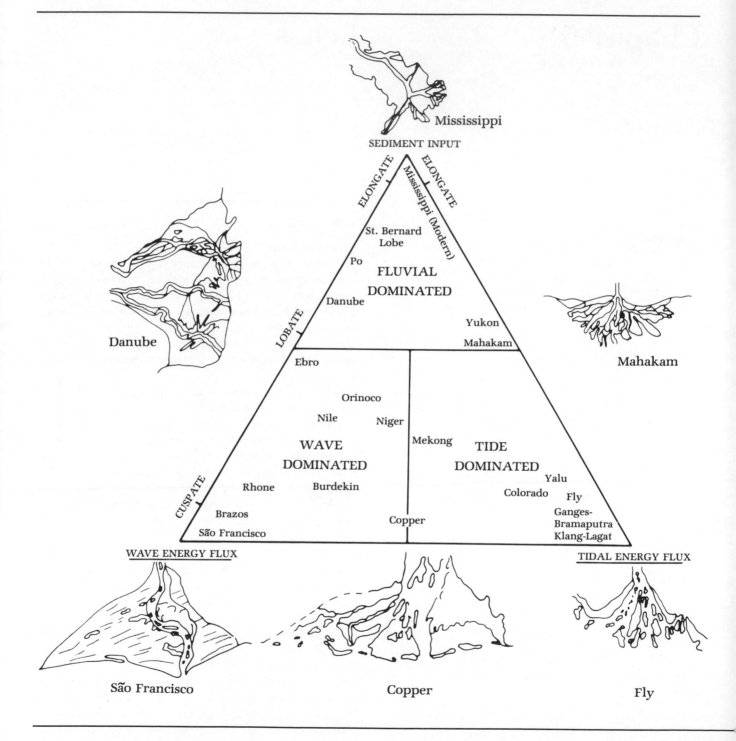

Mississippi

SEDIMENT INPUT

ELONGATE

ELONGATE

Mississippi (Modern)

St. Bernard Lobe

Po

FLUVIAL DOMINATED

Danube

Yukon

Mahakam

LOBATE

Ebro

Orinoco

Nile

Niger

Mekong

WAVE DOMINATED

TIDE DOMINATED

Yalu

Rhone

Burdekin

Colorado

Fly

CUSPATE

Brazos

Copper

Ganges-Bramaputra

São Francisco

Klang-Lagat

WAVE ENERGY FLUX

TIDAL ENERGY FLUX

Danube

Mahakam

São Francisco

Copper

Fly

Mahakam (river-tide). The Niger Delta of west Africa represents a balanced deltaic system, characterized by a nearly equal intensity of wave- and tidal-energy flux, and a moderate to strong fluvial component.

DEPOSITIONAL HYDRAULICS OF RIVER-DOMINATED DELTAS

In addition to river discharge, wave action (and associated longshore currents), tidal action, and turbidity currents may also be significant in deltaic deposition. In river-dominated deltas, the density contrast between the river effluent on one hand, and the oceanic standing body of water into which the river builds a delta will control the type of delta that is developed. The rational approach used to explain the hydraulics of delta formation is an application of engineering theory of jet flow (Bates, 1953). This theory assumes that river inflow into the ocean is comparable to the discharge of a turbulent jet through a well-defined stable orifice (Figure 5.2). The jet flow is either two-dimensional or three-dimensional, referred to as plane jet and axial jet, respectively. Three types of jet flow are recognized, according to the differences in density contrast between the jet flow itself and the ambient standing body of water into which the jet flows (Figure 5.2).

Bates (1953) recognized three types of jet flow styles that were germane to deltaic sedimentation. These are hyperpycnal jet inflow, homopycnal jet inflow, and hypopycnal jet inflow.

HYPERPYCNAL INFLOW

Hyperpycnal inflow is a plane jet, oriented vertically. The inflowing river water is more dense than the surrounding oceanic or lake water. The jet flow moves along the ocean floor, acquires rapid acceleration, erodes material on the delta front, cuts delta front troughs (or submarine canyons), and deposits sediment at the foot of the delta where gradients decline. Expressed in another way, hyperpycnal jet inflow is the same as a turbidity current. Its develop-

FIGURE 5.1. *Ternary classification of deltas according to dominant depositional mode (from Galloway, 1975; republished with permission of the Houston Geological Society).*

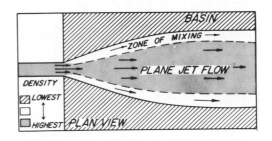

(A) HYPERPYCNAL INFLOW

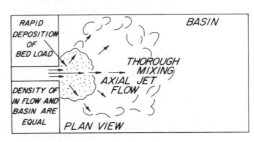

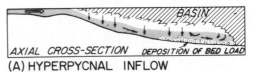

(B) HOMOPYCNAL INFLOW

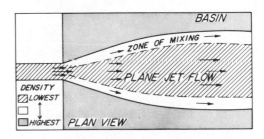

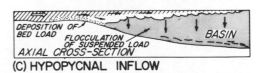

(C) HYPOPYCNAL INFLOW

FIGURE 5.2. *Effluent jet flow models for deltaic systems (after Bates, 1953).*

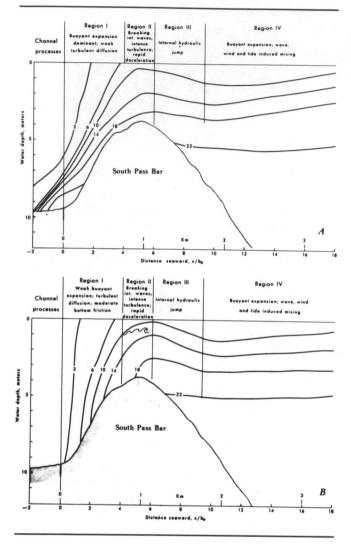

FIGURE 5.3. *Density cross-sections of flow systems at South Pass, Mississippi Delta during rising river stage, 31 January 1973; (A) Flood tide; (B) Ebb tide (from Wright and Coleman, 1974; republished with permission of the University of Chicago Press).*

ment is favored particularly during periods of peak flooding of major rivers, and it may account for some of the interbedded sand lithologies in delta front sediments. Hyperpycnal jet inflows were observed in Lake Geneva on the Rhone Delta by Houbolt and Jonkers (1968) following increased sediment yield and river discharge following spring thaws in the Swiss Alps.

HOMOPYCNAL INFLOW

Homopycnal inflow is an axial jet, where the density of the inflowing river is almost equal to that of the ambient water of an ocean or lake. It is thought to be common to lakes (Bates, 1953). Supposedly, this form of jet inflow produces the so-called Gilbert Delta consisting of topset, foreset, and bottomset beds. Apparently, this type of delta is preserved rarely and only one excellent case of such an ancient lacustrine delta has been documented by Stanley and Surdam (1978).

HYPOPYCNAL INFLOW

Hypopycnal jet inflow is a plane jet oriented in a horizontal plane. The inflowing river effluent is less dense than the surrounding ocean water into which it flows. The ocean water is more dense because of its salinity and its colder temperature. Tidal currents tend to amplify the density contrast at the river mouth by the intrusion of a wedge of saline water into the river mouth. The sea water then buoys up the river water and its suspended sediment content. Tidal action and wave action flush this suspended clay into interdistributary bays, although some of the clay is also deposited by suspension settlement onto the ocean floor. The type of delta that is generated by this process is an imbricating delta, consisting of a series of imbricated deltaic lobes. The delta of the Mississippi River shows such an imbricate history (Scruton, 1960; Kolb and Van Lopik, 1966; Coleman, 1980; Roberts, Adams, and Cunningham, 1980).

The changes in flow conditions associated with hypopycnal jet inflow at the mouth of the Mississippi River were observed in a detailed study by Wright and Coleman (1974). There, they observed that the size and areal dimension of the effluent jet flow change with each phase of the tidal cycle. The density stratification was also observed to change during each phase of the tidal cycle. In particular, during the ebb phase, the zone of maximum

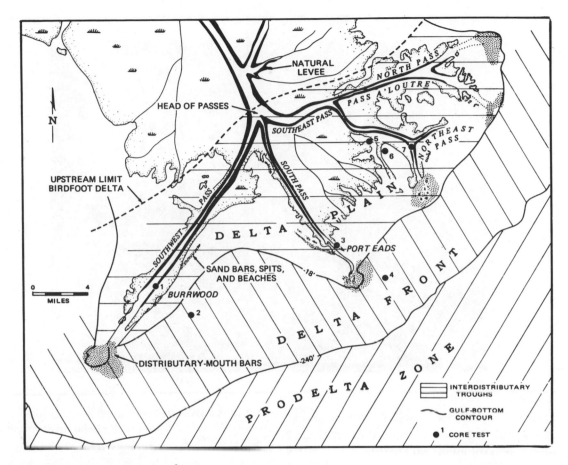

FIGURE 5.4. *Depositional environments and subenvironments, Mississippi Birdfoot Delta (from Gould, 1970; after Fisk, 1961; republished with permission of the Society of Economic Paleontologists and Mineralogists).*

density contrast tends to impinge on the distributary mouth bar of each river mouth, particularly during periods of peak flooding (Figure 5.3). Once the effluent jet flows seaward over the bar, it was observed to expand and lose velocity and hydraulic effectiveness. The jet flow was observed also to expand in this seaward zone. One of the consequences of this impingement of the density interface over the bar crest, combined with the advective effect of breaking internal waves at the density interface, is that the crest of the distributary mouth bar is reworked extensively, giving enhanced porosity and permeability to the sands at the bar crest. Thus, the bar-crest zone be-

comes one of the most favorable targets in exploration for deltaic sandstone stratigraphic traps.

MISSISSIPPI DELTA (RIVER-DOMINATED)

The delta of the Mississippi River is the type example of a river-dominated delta (Coleman, 1976, 1980). This delta is also one that has been studied in the most detail compared with other deltas in the world, and as a consequence, it has been used as a reference standard, perhaps a little too widely. Because of the extensive data published on the Mississippi Delta, geologists tend to overlook deltaic variability when examining ancient counterparts. The discussion that follows is taken from Fisk et al. (1954), Fisk (1956, 1961), Gould (1970), Morgan, Coleman, and Gagliano (1968), Coleman and Gagliano (1965), Huang and Goodell (1970), Scruton (1956, 1960),

Wright and Coleman (1974), Coleman (1976), Prior and Coleman (1977, 1978), Prior and Suhayda (1979), and Roberts, Adams, and Cunningham (1980).

FIGURE 5.5. *Aerial view of small crevasse-splay fan in Garden Island Bay, Mississippi Delta (photograph taken in November 1967).*

The Mississippi Delta is organized into a series of subcomponent environments (Figure 5.4). The most landward zone is an alluvial plain characterized by meandering channels, levees, and a floodplain that contains dominant swamp vegetation. This subenvironment grades into the upper delta plain, which is the nonmarine portion of the delta plain itself. Within the upper delta plain, channel floors are the site of sand transport, grading laterally into silty levees and an interchannel zone of mud and marsh. The lower delta plain is characterized by a breakup of the main river channel into a series of distributary channels. The channels diverge from the main channel and are bounded by silty levees that separate the channels from interdistributary bays. The channels transport sand dominantly as bedload, whereas levee aggradation occurs by bankfull and overbank deposition of fine sand mixed with silt and clay. Mud is transported down the channels as suspended load; this suspended load is distributed and deposited into the interdistributary bay as overbank sediment, as sediment diverging from the channel through crevasse splays or through a more complex route as river effluent, which is then redistributed by waves into the interdistributary bays. The process of crevassing, or erosion of levees, causes some of the riverine flow to be diverted into the interdistributary bay environment and leads to the deposition of a small crevasse-splay fan (Figure 5.5). Marsh deposition is also common to the interdistributary bay environment.

Sand is deposited as distributary mouth bars (Figure 5.6) at the river mouth. It accumulates there because of the dumping action of sand where the river enters the ocean and hypopycnal jet inflow becomes operative. These sands are then reworked and reshaped by a com-

FIGURE 5.6. *Aerial view of distributary mouth bar at Southwest Pass, Mississippi Delta. Bar was exposed during unusual spring low tide, April 1969.*

bination of wave action, longshore currents, and the breaking of internal waves at the interface of suspension sediment-rich effluent and colder salt water (Wright and Coleman, 1974). This reworking is increased during seasonal flooding, enhancing porosity in the crestal zone. Longshore currents elongate the bars parallel to depositional strike. However, it is important to remember that distributary bar growth appears to be a continuous process (Fisk, 1961; Gould, 1970); as the channels are elongated, they are elongated by eroding into the central part of the distributary mouth bar crest and the distributaries are then elongated seaward by additional bedload deposition (Figure 5.7) and progradation. As a consequence, the bars are preserved as linear, bifurcating bar fingers (Figure 5.8). With continued progradation, the bar fingers are buried by channel fills of mud (when the channels are abandoned) and levee and interdistributary bay sediments. Thus, the bar fingers appear to lace their way

through interdistributary bay muds, levees, and delta front sediments (Figure 5.9).

It must be reemphasized that the distributary mouth bars show a distinct zonation of sediment texture, sedimentary structures, and bedding types. These are controlled directly by the jet flow processes at the river mouth and over the bar (Wright and Coleman, 1974). As shown in Figure 5.10, the back-bar sediments are poorly sorted because they are deposited in a zone of buoyant expansion of jet flow. The crest zone is the best sorted because it accumulates in a zone of breaking internal waves and is reworked by surface ocean waves. The bar front sands are interbedded with mud and silt because they accumulate in a zone of expanded jet flow that undergoes deceleration and a hydraulic jump from upper to lower flow regime flow. The distal bar zone tends to be poorly sorted because it accumulates in a zone of buoyant expansion of jet flow enriched in silts and clays.

From the point of view of an exploration geologist, the obvious candidate for excellent reservoir sands should be in the bar-crestal zone. However, the explorationist is warned that the central portion of the crest where one

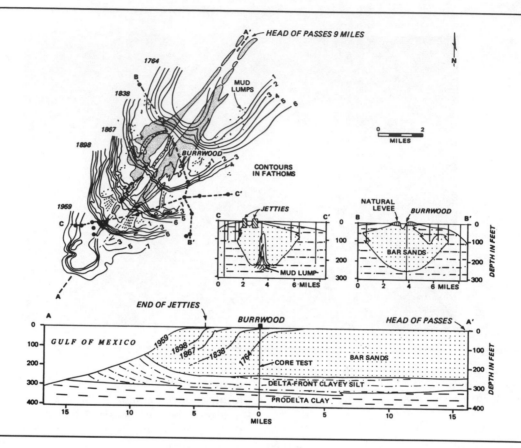

FIGURE 5.7. *Historic development of progradation of Southwest Pass bar finger, Mississippi Delta (from Gould, 1970; after Fisk, 1961; republished with permission of the Society of Economic Paleontologists and Mineralogists).*

would expect the thickest sandstone isopach (Figure 5.8) is likely to be the site of channelization by extension of the distributary channel during delta growth. As a consequence, a well drilled in the zone of the potentially thickest isopach is, most likely, going to penetrate channel-filled clays deposited during channel abandonment. It is advised strongly that wells be drilled offset from this axial mud-channel-fill zone to penetrate the levee or inner distributary bay and then penetrate the crestal zone of the distributary mouth bar, which would make an excellent petroleum reservoir rock. The channel mud fill, in addition to the levee and bay sediments, makes an excellent impermeable seal.

Seaward of the distributary mouth bars is the delta front consisting of interbedded sands and muds. Some of these sands are graded with sharp basal scours and gra-

dational tops (Coleman, 1980) and it is suggested that some of these are deposited as turbidites during seasonal spring or longer-term floods when the river mouth supposedly changes from hypopycnal to hyperpycnal jet inflow.

The prodelta zone, which occurs seaward of the delta front zone, is dominated by mud deposition from suspension processes.

The growth and development of the Mississippi Delta alternates between constructional and destructional phases (Scruton, 1960). During the constructional phase, delta progradation toward the basin center proceeds by

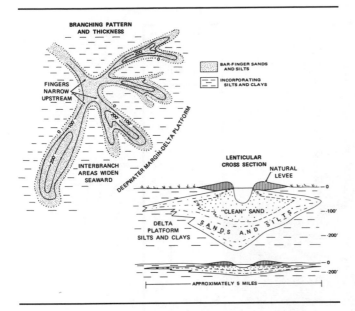

FIGURE 5.8. *Geometry and sedimentary characteristics of Mississippi Delta bar finger sands (from Gould, 1970; after Fisk, 1961; republished with permission of the Society of Economic Paleontologists and Mineralogists).*

means of seaward progradation of distributary channel systems and associated distributary mouth bars (Figure 5.7), which averages 80 m/yr. As these channel and bar systems build out seaward by bedload deposition, suspended load is diverted from the channels either by wave and tidal action or diversion through crevasse splays into the interdistributary bays. Some of this sediment is also dispersed into the bays and marshes by storms (Baumann, Day, and Miller, 1984). Consequently, the interdistributary bays will prograde seaward also, but at a slower rate. Similarly, some of this suspended load also is deposited in the delta front and prodelta zones. Evidence for such progradation includes observing pairs of identical contours on bathymetric maps surveyed at different times; the more recent maps show contours displaced seaward compared with the older maps (Scruton, 1960). Second, navigation charts show a history of continued progradation of distributary mouth bars through time, as shown in Figure 5.7.

As the channel systems that provide sediment are elon-

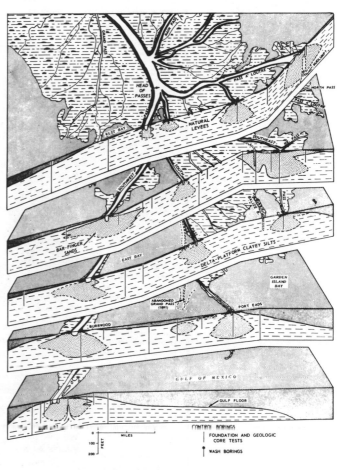

FIGURE 5.9. *Model of Mississippi Birdfoot Delta showing geometry of distributary channel systems (from Fisk, 1961; republished with permission of the American Association of Petroleum Geologists).*

gated basinward, the efficiency of these channels for sediment transport decreases. With time, such channels will shift and divert channel flow to the ocean by a more efficient and shorter route. This process of channel shifting leads to development of smaller delta lobes, as well as the larger imbricated deltaic units in southern Louisiana. Such shifts lead to the deposition of new deltas (Scruton, 1960; Coleman, 1980; Roberts, Adams, and Cunningham,

98

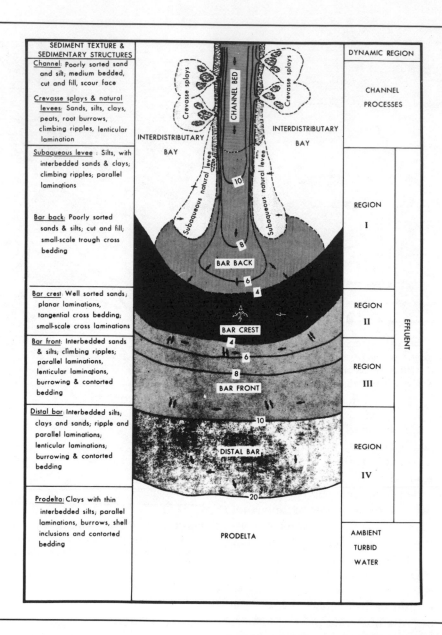

FIGURE 5.10. *Morphological and sedimentary characteristics of a stratified river mouth and relationships to effluent dynamic regions (from Wright and Coleman, 1974; republished with permission of the University of Chicago Press.)*

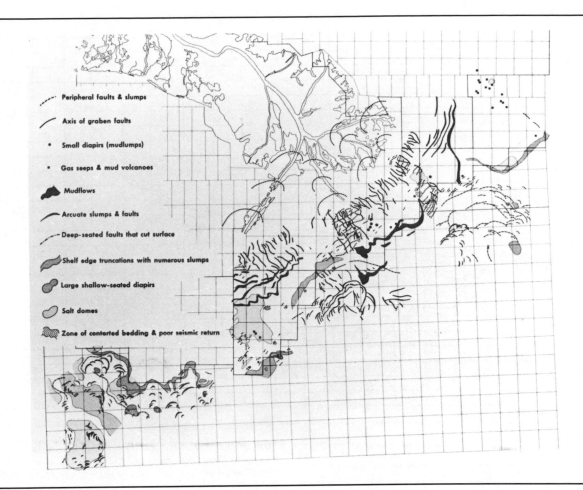

Peripheral faults & slumps

Axis of graben faults

Small diapirs (mudlumps)

Gas seeps & mud volcanoes

Mudflows

Arcuate slumps & faults

Deep-seated faults that cut surface

Shelf edge truncations with numerous slumps

Large shallow-seated diapirs

Salt domes

Zone of contorted bedding & poor seismic return

FIGURE 5.11. *Distribution of subaqueous faults and sediment gravity flows, Mississippi Delta (from Coleman, 1976; republished with permission of the CEPCO Division, Burgess Publishing Co.).*

1980) and result in a composite delta consisting of a series of lenticular packets.

Following channel shifting, the abandoned delta undergoes a destructive phase as riverine sediment yield is diverted. The older distributary channels fill in slowly with fine sediments and vestiges of the original channels merge into the deltaic landscape. The abandoned delta also subsides below sea level, and once it does so, transgressive, wave-dominated processes become dominant. Consequently, the sands on the edge of the abandoned delta are reworked by waves and redistributed as a thin sand veneer that migrates landward and accumulates as a chain of barrier islands (for example, Chandeleur Islands). Landward of these transgressive barriers, lagoonal clays with marine fauna will accumulate. Because of the large rate of sedimentation associated with the prior constructional phase, the delta plain of the progradational portion of the delta is then overlain by a marine clay or a marine transgressive barrier sand.

Superimposed on both the constructional and destructional histories of the Mississippi Delta are two other modifications. The first of these is the tectonic development of mud-lump diapirs. As shown by Morgan, Coleman, and Gagliano (1968), distributary channel systems

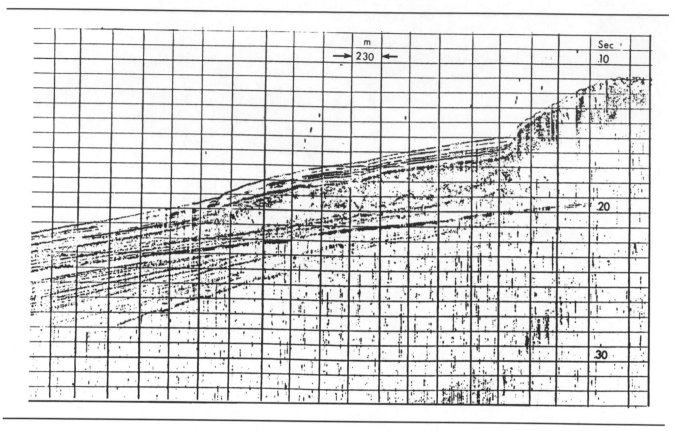

FIGURE 5.12. *Seismic record across a subaqueous mudflow or debris flow, Mississippi Delta (from Coleman, 1976; republished with permission of the CEPCO Division, Burgess Publishing Co.).*

extend seaward and build bar finger sands, and the net weight of the sands causes water-saturated clays of the delta front to migrate from areas of greater overburden pressure to least overburden pressure. This migration is in the form of a plastic flow and on reaching zones of least stress, the clays will warp upward into a fold that is later breached by a high-angle thrust fault. The clay, which is now more coherent, moves upward along such a high-angle reverse fault and intrudes overlying sediment as a diapir, or is intruded above sea level and appears as a mud-lump island. The timing of appearances of mud-lump islands (recorded on navigation charts) was correlated to additional sand deposition on distributary bars and seaward progradation of such bars (Morgan, Coleman, and Gagliano, 1968).

A second set of modifications owes its origin to oversteepened slopes and downslope gravity processes. As shown by Coleman (1976), Prior and Coleman (1977,

1978), and Prior and Suhayda (1979), several types of subaqueous gravity processes occur on delta fronts of the Mississippi (Figure 5.11). These include peripheral faults, slumping, graben faulting, and mudflows. Evidence for extensive slumping, mudflows (Figure 5.12), and graben faulting (Figure 5.13) includes both seismic records and side-scan sonar mapping. The triggering mechanism for these features appears to be oversteepening of water-saturated delta front sediments. Slumping of distributary bar sands is favored following deposition after annual flooding; thick water-saturated bar sands tend to fail an average of four years after deposition by oversteepening of slopes in response to diapirism, ma-

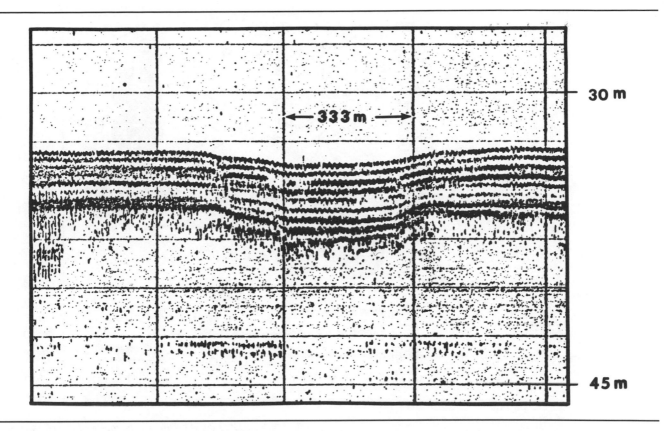

FIGURE 5.13. *Seismic profile over radial graben system, southeast of South Pass distributary, Mississippi Delta (from Coleman, 1976; republished with permission of the CEPCO Division, Burgess Publishing Co.).*

jor storms and hurricanes, or increased pore-pressure caused by generation of biogenic gas (Lindsay, Prior, and Coleman, 1984). Coleman (1976) showed seismic sections and side-scan sonar records for all these features and the reader is encouraged to examine those for additional details. Similar slumps were reported by Shepard (1973) from the Magadalena Delta, Columbia, and by Klein, DeMelo, and Della Favera (1972) in an ancient example in the Cretaceous of the Reconcavo Basin, Brazil.

NIGER DELTA

Data covering the Niger Delta includes work by Allen (1965b, 1970), Burke (1972), Oomkens (1974), and Weber (1971). The Niger Delta contains a large sand content

dispersed from the Niger River, which possesses an annual discharge rate of 2×10^{11} m³. The wave-energy flux, generated from a large area of fetch over the South Atlantic, is moderate to large. The tidal range is mesotidal, averaging 2.2 m. The external morphology of the delta is symmetrical and fan shaped, and in plan appears more like the head of a pick axe. The delta plain consists of braided distributaries filled with sand, lacing through muds and mangrove swamps. In the lower delta plain, tidal exchange becomes more significant, and sand-filled tidal channels cut through the mangrove swamps.

The seaward edge of the delta plain consists of a series of prograding barrier islands cut by tidal channels and tidal inlets (Figure 5.14). These barriers consist of sand and are built by a combination of longshore current and inlet migration. The tidal channels cutting the barriers are filled with sand, and provide sand on the seaward edge, which accumulates according to Allen (1965) as

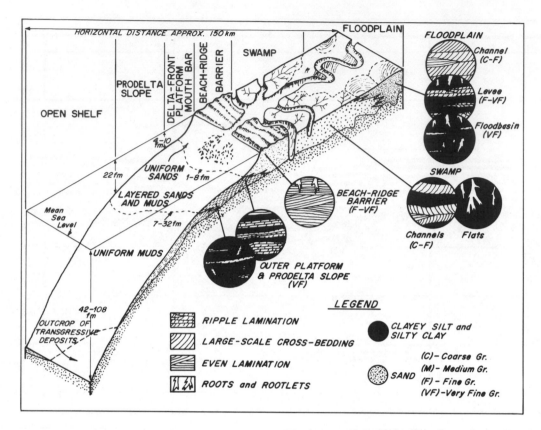

FIGURE 5.14. *Principal sedimentary facies of the Holocene Niger Delta (redrawn after Allen, 1970).*

"mouth bars" but these are, in reality, ebb-tide-dominated tidal deltas. The prodelta and the delta front are similar to the Mississippi Delta (Figure 5.14).

The facies profile through the delta (Figure 5.14) indicates that in time, this delta progrades seaward and generates a coarsening-upward sequence (see also Weber, 1971; Oomkens, 1974). This vertical sequence would consist of basal prodelta clays, overlain by delta front silts and sands grading upward into tidal delta sands, overlain by barrier sands grading laterally into tidal-channel fills (sand) and delta plain mangrove swamps. The delta plain facies that would be preserved would differ in style from the Mississippi.

The Niger Delta builds over a very narrow shelf region. Therefore, it has been suggested that at the present time and in the past, the Niger Delta would prograde over submarine fans (Figure 5.15) at the foot of the continental slope (Burke, 1972). This relationship is significant for

petroleum exploration because it suggests that if one exploits such types of deltas, it should be possible to drill a single well to extract oil and gas from multiple pay zones in the delta plain zone at the top of a delta sequence, and deep-water submarine fan sands underlying the delta sequence (Figure 5.15).

TIDE-DOMINATED DELTAS

The two best-documented tide-dominated deltas are the Klang-Langat Delta of Malaysia (Coleman, 1976; Coleman, Gagliano, and Smith, 1970) and the Ord River Delta of northwest Australia (Coleman, 1976).

KLANG-LANGAT DELTA

This delta is formed by the merging of the Klang and Langat rivers. River and sediment discharges are small, as is wave action. The tidal range fluctuates from mesotidal to macrotidal during spring tides (Figure 5.16). The delta plain consists of tidal flats, mangrove swamps, and two-way flowing tidal channels that are counterparts to the distributary channels of the Mississippi Delta. Levees are absent. The tidal flats are dominantly areas of mud accumulation, whereas the two-way tidal-channel systems are filled with sand. The delta front consists dominantly of sand, which represents a coalescing of tidal-current sand ridges forming on the seaward edge of the delta in a manner similar to those described from the North Sea (Houbolt, 1968) and the Yellow Sea (Klein et al., 1982). The prodelta zone seaward of these tidal sand bodies consists of clay.

The vertical sequence of this delta consists of basal muds of the prodelta zone, tidal sand bodies of the delta front zone, and a delta plain of tidal mudflats cut by two-way tidal channel sands overlain by tidal flats and mangrove swamps.

ORD RIVER DELTA

The Ord River Delta occurs on the northwest coast of Australia where the climate is arid. The river discharge of the Ord is relatively small. Delta plain deposits consist of tidal mudflats, intertidal sand bodies and shoals, subtidal sand bodies and shoals, and high intertidal to supratidal mudflats with evaporite minerals. The distal bar, delta front, and prodelta zone appear similar, in terms of sediment types, to the Mississippi River.

WAVE-DOMINATED DELTAS

Coleman (1976) summarized the characteristics of two wave-dominated deltas, the Sao Francisco Delta of eastern Brazil, and the Senegal Delta of Senegal, West Africa. Both are built by a combination of sediment dispersal from riverine sediment yield and reworking by extensive wave action. Tidal range is microtidal and, apparently, tidal currents play a minor role in sediment dispersal.

In both the Sao Francisco Delta and the Senegal Delta, the delta plain consists of a series of beach ridges and coastal dune facies consisting of sand. These are cut by the channel systems of the Sao Francisco River and Senegal River, respectively. Coastal barrier development is enhanced by strong longshore current transport, particularly in the case of the Senegal Delta. The Senegal Delta

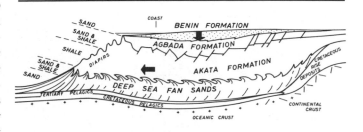

FIGURE 5.15. *Cross-section of Niger Delta, showing deltaic units of prodelta (Akata Formation), delta front (Agbada Formation), and delta plain (Benin Formation) prograding over submarine fan system (from Burke, 1972; republished with permission of the American Association of Petroleum Geologists).*

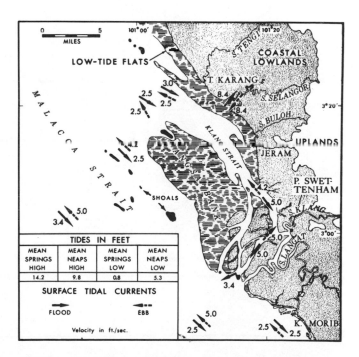

FIGURE 5.16. *Klang-Langat Delta, major physiographic zones, environments, tidal ranges, and tidal currents (from Coleman, Gagliano, and Smith, 1970; republished with permission of the Society of Economic Paleontologists and Mineralogists).*

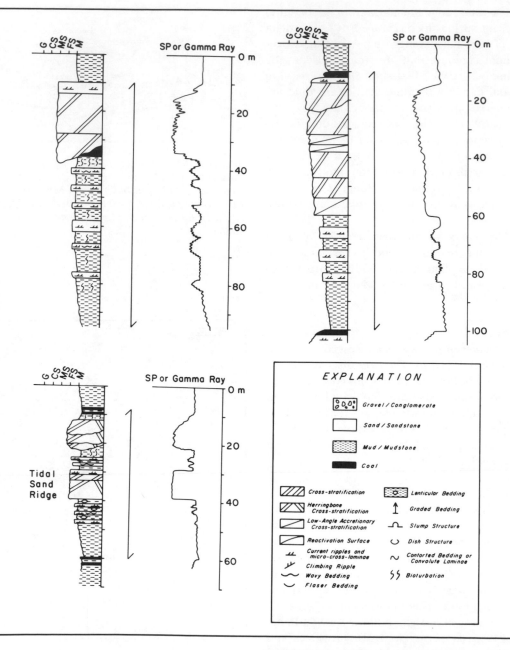

FIGURE 5.17. *Idealized vertical sequences and log motif of river-dominated delta (upper left), wave-dominated delta (upper right), and tide-dominated delta (lower left). Doubly terminated arrows show interval of complete sequence. (Abbreviations: G—Gravel; CS—Coarse sand; MS—Medium sand; FS—Fine sand; M—Mud.)*

shows the development of extensive marsh deposits in the swales between the beach ridges of the prograding barrier island delta plain zone.

The vertical sequences of both of these deltas show basal prodelta muds and delta front sands interbedded with mud, like the Mississippi. The delta plain, however, may show lateral intertonguing of distal bars overlain by channels or more likely, the prograding delta plain beach ridge, coastal barrier, and dune complex. The entire sequence coarsens upward, but the preservation of coastal barriers at the top is the distinguishing criterion for distinguishing wave-dominated deltas from tidal- or river-dominated deltas in the subsurface or in outcrop.

DELTAIC VERTICAL SEQUENCES

The vertical sequences of sedimentary structures, grain size, and lithologies of deltas show a coarsening-upward trend (Figure 5.17). Despite differences in the wave-energy flux, tidal-energy flux, and river influx causing differences in sediment facies in the delta plain, all deltas show a coarsening-upward style. Figure 5.17 shows a comparative representation of vertical sequences from these three major types of deltas. The coarsening-upward sequence motif is self-evident in all. The greatest variation in vertical sequence characteristics of these delta types is found in the delta plain, and to a lesser extent in the delta front. Other examples of such sequences are reviewed in Coleman (1976, 1980).

It is noteworthy that the vertical sequence of wave-dominated deltas shows some similarities to the prograding, regressive marine barrier island sequence discussed in Chapter 4. It is worth asking how one would discriminate between the two environments when examining such sequences in outcrop or in the subsurface. The main differences would be in the thickness of the sequence and in overall geometry. The barrier island coarsening-upward sequence would be relatively thin and would comprise either part of a shoe-string geometry or a sheetlike geometry. The wave-dominated deltaic coarsening-upward sequence would be much thicker normally, and comprise part of a sedimentary body that is fan-shaped in plan and lenticular in cross-section. Updip from wave-dominated deltas, one should be able to observe a record of the fluvial systems that provided the sediment for the delta.

ANCIENT DELTAS

Some of the best-documented ancient examples of deltaic deposition are those described by Ferm (1962, 1970), Ferm and Williams (1963, 1964), and Horne et al. (1978) from the Pennsylvanian of the Appalachian Plateau of the eastern USA. There, extremely complex channel systems are known, including channel fills of mud and sand. These deltas were clearly river dominated with thick distributary bar complexes being particularly well preserved. The Pennsylvanian sandstones of the Illinois Basin are also dominated by deltaic sediments (Wanless et al., 1970), although because of their deposition in a cratonic setting, the coarsening-upward deltaic sequences are relatively thin, implying that rates of delta progradation may have been large because these deltas prograded into shallow depths of water (Klein, 1974). These Pennsylvanian deltas in the Illinois Basin also show evidence of preserving both a constructive and destructive phase of deposition.

The Cretaceous sedimentary rocks of the Rocky Mountains also show evidence of a deltaic history. The Dakota Sandstone (Weimer and Land, 1972) shows evidence of extensive deltaic deposition, although locally, tidal-flat deposits are common (MacKenzie, 1972b), indicating some tide-dominated modification. The Parkman Sandstone of Wyoming (Hubert, Butera, and Rice, 1972) shows many of the features reviewed above, including some evidence of local slumping. The Frontier Formation of Wyoming (Figure 5.18) shows well-developed coarsening-upward sequences of a river-dominated deltaic system. Laterally, particularly in southwestern Wyoming, these deltas change into wave-dominated deltas (near Kemerer, Wyoming), and to barrier islands (Ryer, 1977).

The Middle Ecca Shale (Permian) of South Africa was also characterized by a deltaic history (Hobday and Matthew, 1975). There, extensive coal sequences comprise a river-dominated delta, particularly in the upper delta plain facies of northern Natal province. In east central Natal, marine deltas are more common. The deltas of the Middle Ecca Shale show well-preserved distributary mouth bar features (Figure 5.19) at the top of coarsening-upward sequences that are identical to the Mississippi Delta.

Other well-known deltaic systems have been recorded from the Pennsylvanian Atoka Formation and Bluejacket

FIGURE 5.18. *Coarsening-upward sequence produced by prograding delta, Frontier Formation (Cretaceous), North Tisdale, Wyoming. (1) Delta plain distributary sandstone; (2) Delta front; (3) Prodelta (from Klein, 1974; republished with permission of the Geological Society of America).*

FIGURE 5.19. *Middle Ecca Shale (Permian), Zinguin Mountain, Natal, South Africa, showing delta front lithologies (striped), overlain by distal toe of imbricate-bedded distributary mouth bar (center right) overlain by delta plain sediments.*

Sandstone of Oklahoma (Visher, Saitta, and Phillips, 1971), the Abottsham Formation (DeRaaf, Reading, and Walker, 1965) of Carboniferous Age of Britain, the Wilcox Group (Eocene) of Texas (Fisher and McGowen, 1969; Edwards, 1981), and the Frio Formation (Oligocene–Miocene) of Texas (Galloway, Hobday, and Magara, 1982). Obviously, the brief summary above is a partial list and many others are described in Morgan (1970), Broussard (1975), and Shirley (1966), among others.

Recently released data from the Baltimore Canyon Trough (Schlee, 1981; Libby-French, 1984) indicates that the outer continental shelf of the Atlantic coast of the USA contains several deltaic units. Libby-French (1984) reported many features such as coarsening-upward log signatures associated with lithologic changes to indicate deltaic and delta front deposition in the Mohawk Sandstone and MicMac Shale equivalents of Upper Jurassic and Lower Cretaceous Age. Schlee (1981) summarized a detailed study of seismic stratigraphy of the Baltimore Canyon Trough also, and demonstrated many deltaic systems characterized by a sigmoidal seismic signature (Figure 5.20).

SOME OIL FIELD EXAMPLES IN DELTAS

Deltas represent one of the four major environments of deposition from which oil and gas are produced and exploited. Obviously, the many examples that could be cited range beyond the scope of this book, so it is intended here to focus on a few key problems and a few examples.

Before reviewing some specific cases, it seems appropriate to outline a procedure of subsurface facies analysis in deltas. Earlier, it was demonstrated that by combining analysis of the nature of gamma-ray and SP log shapes with the presence or absence of carbonaceous detritus and glauconite (Selley, 1976), it is possible to interpret depositional settings in the subsurface (Figure 1.1).

The deltaic environment represents a complex set of subcomponent depositional settings that experience a constructional and destructional history. Each of these depositional settings is characterized by its own SP and gamma-ray log characteristics that are recognizable in the subsurface, permitting easy interpretation of the history of a specific delta lobe, and enabling the petroleum geologist to predict new step-out wells along with new

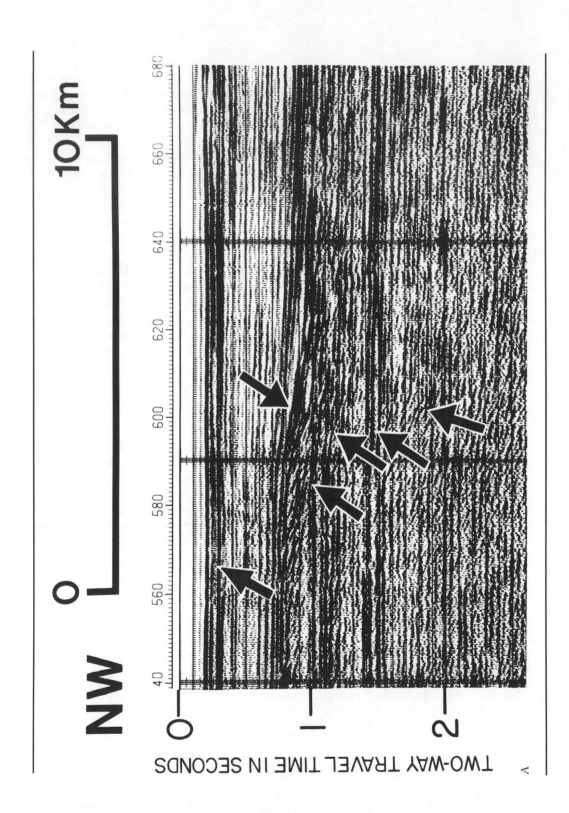

NW

10Km

0

TWO-WAY TRAVEL TIME IN SECONDS

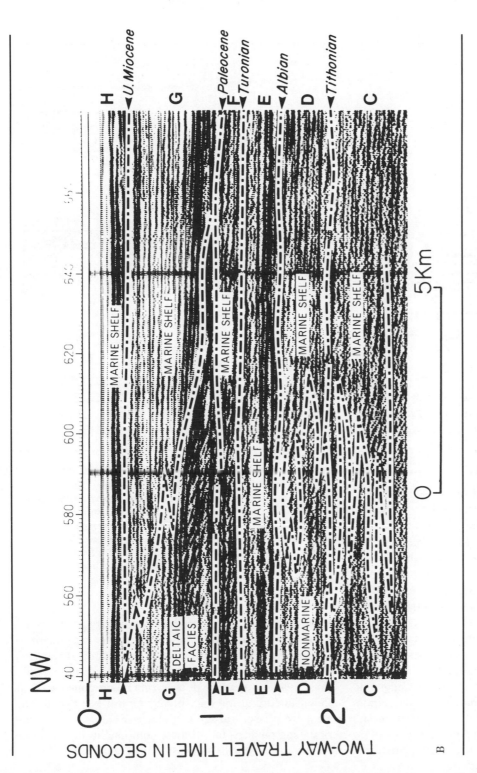

FIGURE 5.20. (A) *Seismic profile, outer continental shelf, southern New Jersey with arrows indicating conspicuous reflectors;* (B) *Paleoenvironmental interpretation of profile shown in A, with shingled deltaic signature (seismic section photographs courtesy of J.S. Schlee; from Schlee, 1981; republished with permission of the American Association of Petroleum Geologists).*

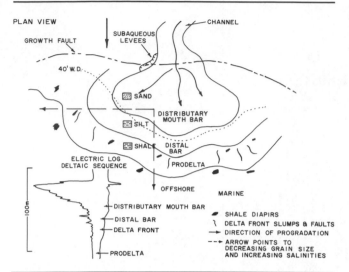

FIGURE 5.21. *Plan view and diagnostic electric log motif for river-dominated delta with distributary mouth bar system (redrawn after Saxena, 1976a,b).*

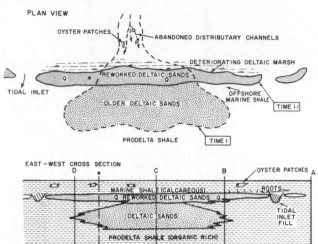

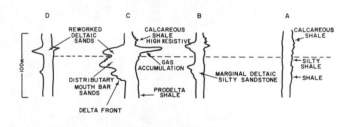

FIGURE 5.22. *Map view, cross-section, and electric log motif through delta destructive system. In plan view, Time I represents maximum constructional stage of distributary bar progradation, whereas Time II represents a subsequent destructive stage with wave-dominated barrier transgressive sand overlying distributary. Cross-section shows both constructional and destructional stages, with lower panel showing electric log features at specific places marked in cross-section (redrawn after Saxena, 1976a,b,c).*

locations (Saxena, 1976a,b,c). The best summary of how to approach the subsurface analysis of deltaic depositional systems was presented by Saxena (1976a,b,c) and is based both on experience with Holocene deltas (primarily the Mississippi) and several unpublished exploration tests.

The components of deltaic environments, their constructional and destructional history, and their tectonic and slope modifications were reviewed earlier. The best targets for exploration for oil and gas in deltas are the distributary mouth bar sand complexes and the destructive-phase, transgressive barrier sands. Delta building (constructional stage) causes the deposition of distributary mouth bar sands that are characterized by a specific electric log pattern (Figure 5.21), reflecting a coarsening-upward nature. When a delta undergoes a channel shift, the delta is abandoned and subjected to a destructive history (Scruton, 1960; Coleman, 1980). This destructive history includes wave modification of the delta edge and the development of transgressive barrier sands (Figure 5.22) overlying the earlier-deposited distributary mouth bar sands. The characteristic electric and gamma-ray log patterns that would be recorded in a well penetrating

sands that had experienced such a history are shown in Figures 5.22 and 5.23. The characteristic log pattern for upper delta plain meandering channel sands is shown in Figure 5.24. Both such channels, as well as crevasse-splay fans, may be deposited over other components of the delta, including the distributary mouth bar and the destructive-phase, transgressive barrier sands. A summary of the log patterns for all these components of a delta are shown in part, or in succession, in Figure 5.25.

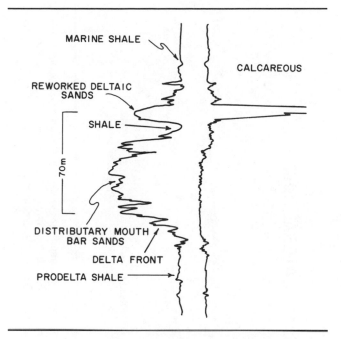

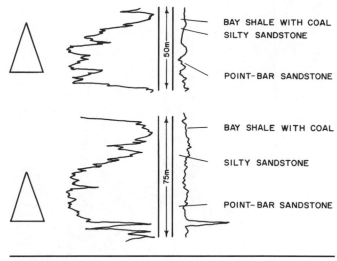

FIGURE 5.24. *Electric log motif for two meandering alluvial valley point-bar sequences (redrawn after Saxena, 1976c).*

FIGURE 5.23. *Idealized electric log pattern of a delta destructive system (redrawn after Saxena, 1976c).*

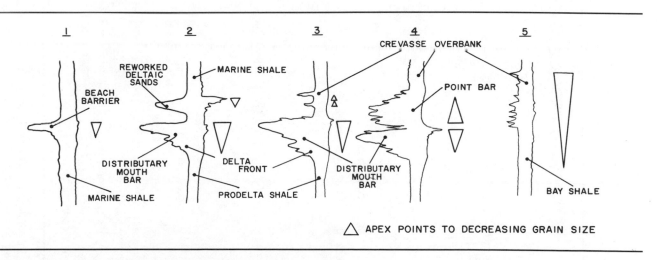

FIGURE 5.25. *Series of electric log motifs showing (1) beach barrier, (2) delta destructive system, (3) distributary overlain by crevasse-splay fan, (4) distributary overlain by point bar, and (5) a series of crevasse-splay fans in bay sequence (redrawn after Saxena, 1976c).*

SOUTHWARD PROGRADATION AND LATERAL SHIFT

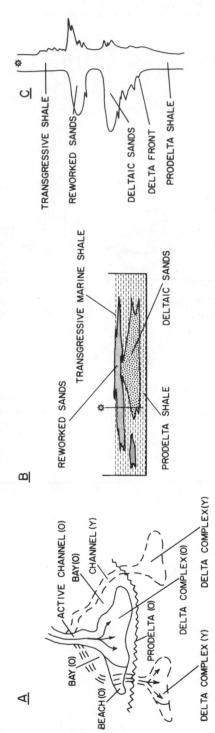

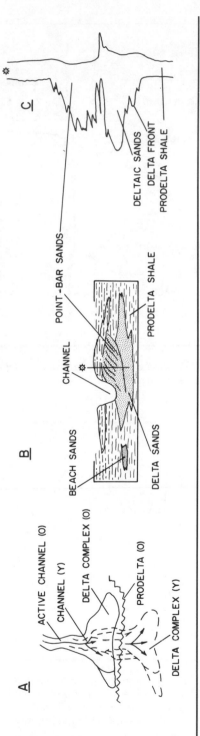

SOUTHWARD PROGRADATION; NO LATERAL SHIFT

FIGURE 5.26. Upper. *Suggested electric log sequence representing southward progradation and lateral shifting leading to development of delta-destructive system. (A) Plan, (B) section, and (C) electric log (redrawn after Saxena, 1976c.) Lower. Suggested electric log sequence showing history of southward progradation of delta system, without lateral shift. (A) Plan, (B) section, and (C) electric log (redrawn after Saxena, 1976c.)*

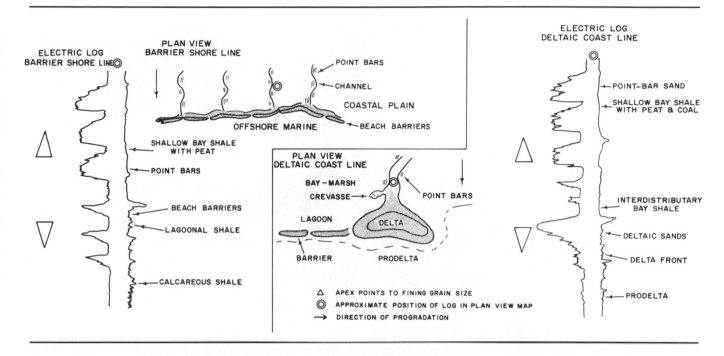

FIGURE 5.27. Left. *Electric log from well shown in map with double circle, indicating history of prograding barrier systems overlain by series of meandering channels (redrawn after Saxena, 1976c.)* Right. *Electric log from well shown in map with double circle, indicating a history of a distributary mouth bar being channelized and overlain by meandering point-bar sands (redrawn after Saxena, 1976c.)*

According to Saxena (1976a,b,c), other examples exist that permit one to reconstruct deltaic shifts assuming a sequence of certain depositional events. Figure 5.26 (upper panel) shows the SP log pattern that develops when a well penetrates a sequence of sediments of delta distributary abandonment followed by a transgressive destructive phase and later deposition of a transgressive barrier bar analogous to the Chandeleur Islands. The lower panel of Figure 5.26 shows the log sequence if a well penetrates sands where delta progradation was followed by a seaward extension of the meandering alluvial upper delta plain over the distributary mouth bar. Other combinations of depositional events and their electric log characteristics are shown in Figure 5.27 where both channel progradation over barrier island systems (left),

and channel progradation over a distributary mouth bar (right), are shown. Other combinations, such as splay over distributaries, or splay over transgressive barriers, can also be considered and the electric log pattern that should emerge using data in Figures 5.21, 5.22, 5.24, and 5.25 is left as an exercise to be undertaken by the reader.

Among some of the better examples of deltaic stratigraphic traps, the Booch Sandstone of Oklahoma (Busch, 1959) has attracted the most attention. Busch (1959) was able to delineate a system of distributaries from isopach mapping and, as shown in Figure 5.28, the resistivity and SP log patterns of the Booch Sandstone fit the above discussion about subsurface log analysis of deltaic systems. Other Paleozoic deltas from which oil production is known include the Bluejacket system of Oklahoma (Visher, Saitta, and Phillips, 1971), the Strawn Series of Texas (Shannon and Dahl, 1971), and the Cisco Series of Texas (Bloomer, 1977).

The Miocene and Pliocene of the Gulf Coastal Plain are well known for their deltaic sandstone reservoir systems (Clark and Rouse, 1971; Curtis and Picou, 1978; Sangree and Widmier, 1977; Stuart and Caughey, 1977). Deltaic

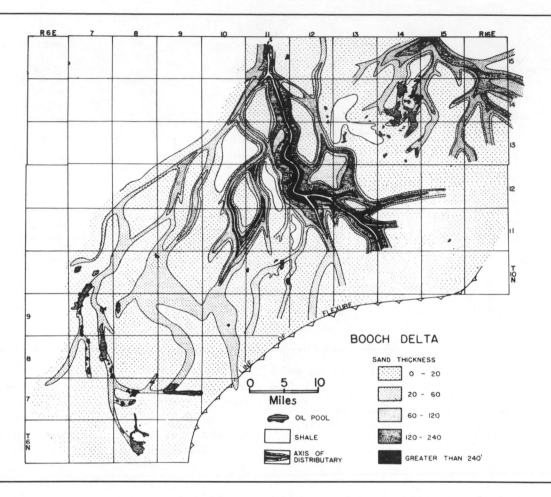

FIGURE 5.28. *Sandstone isopachs defining distributary channel system in Pennsylvanian Booch Sandstones, of deltaic origin, Oklahoma (from Busch, 1959; republished with permission of the American Association of Petroleum Geologists).*

lobes have been delineated (Curtis and Picou, 1978). Such lobate reservoirs appear in seismic sections, as illustrated from the Miocene of California (Figure 5.29). The Eocene Wilcox Group and the Oligocene–Miocene Frio Formation of Texas also contain extensive delta distributary reservoirs (Fisher and McGowen, 1969; Edwards, 1981; Galloway, Hobday, and Magara, 1982). The trapping of oil and gas in the Cenozoic Gulf Coast examples is aided by piercement salt domes and growth faults (Edwards, 1981). Prodelta and interdistributary mudstones provide both source beds and reservoir seals.

The above examples appear to occur mostly from river-dominated deltaic counterparts. Production in the Niger

Delta (Weber, 1971), however, is from a more balanced system, and most of the reservoirs are in the barrier island complexes, and sandy tidal-channel fills that comprise the delta plain of this delta system. There, growth faulting (Weber, 1971; Oomkens, 1974) has also acted to seal part of the reservoir. The Mahakam Delta of Borneo is a mixed tide-dominated and river-dominated delta, and there production appears to come from delta plain tidal-channel systems (Combaz and DeMatharel, 1978).

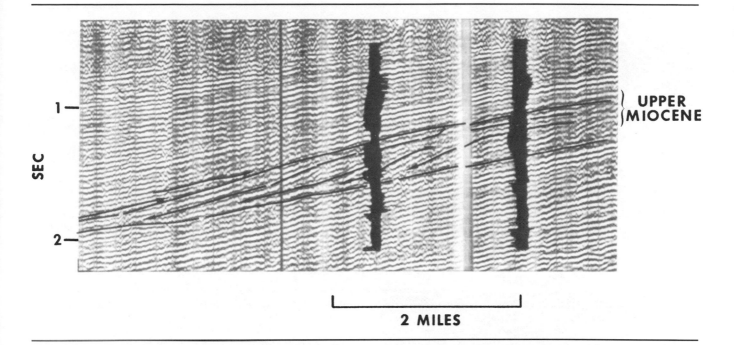

FIGURE 5.29. *Seismic section of oblique-progradational deltaic, seismic facies, Miocene, San Joaquin Valley, California (from Sangree and Widmier, 1977; republished with permission of the American Association of Petroleum Geologists).*

Recent work on the seismic characteristics of deltaic systems by O.R. Berg (1982) has demonstrated that distinctions between river-dominated and wave-dominated deltas can be made from seismic signatures (Figures 5.30 and 5.31). River-dominated deltas tend to be expressed on seismic sections as complex-oblique, tangential-oblique, complex sigmoidal-oblique, and sigmoidal signatures, whereas wave-dominated deltas tend to show a shingled and oblique-parallel pattern (Figures 5.30 and 5.31). Many subsurface deltas overlie and are associated with turbidite systems as well, according to O.R. Berg (1982; Figure 5.32). A comparison of Figure 5.15 from the Niger Delta (Burke, 1972) with Figure 5.32 from O.R. Berg (1982) is of interest and is suggested as an exercise for the reader.

Deltaic hydrocarbon reservoirs have also been observed in conjunction with three-dimensional (3-D) seis-

mic surveys, the best case history being represented in Miocene–Pliocene offshore sediments in the Gulf of Thailand (Brown, Dahm, and Graebner, 1981). There, a series of computer-retrieved horizontal seismic maps (seiscrop section) delineated a series of sand bodies, which are meandering channels, barrier islands in a destructive delta, and offshore bars that are clearly part of a major delta system containing natural gas (Figure 5.33). The 3-D seismic survey method represents a powerful tool in mapping sand bodies simply because machine computation methods permit data retrieval of horizontal seismic data in millisecond intervals, providing the necessary resolution for mapping sand bodies at least 5 m thick.

DELTAIC COAL

Coal occurrences are attributable to deltaic depositional environments. The Pennsylvanian coal fields of the Appalachian Basin (Horne et al., 1978) and of the Illinois Basin (Wanless et al., 1970) are correlated directly to deltaic systems. In the Appalachians, Horne et al. (1978) demonstrated that thin coals are often associated with

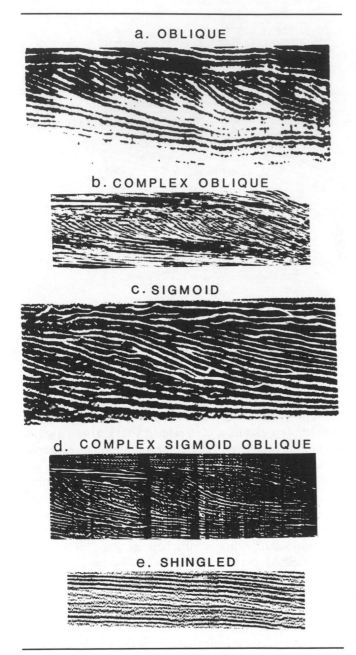

FIGURE 5.30. *Seismic reflection patterns of river-dominated deltas (a, b, c, d) and wave-dominated (e) deltas (photograph of seismic sections courtesy of O.R. Berg; from O.R. Berg, 1982; republished with permission of the American Association of Petroleum Geologists).*

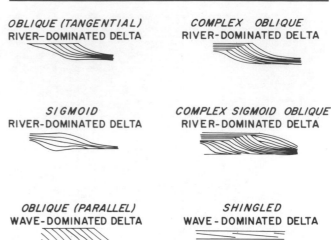

FIGURE 5.31. *Model reflection patterns of river-dominated and wave-dominated deltas (redrawn from O.R. Berg, 1982).*

transgressive barrier facies in back-barrier environments, whereas lower delta plain coals are thin, but widespread. The thickest coal accumulations tend to occur mostly in upper delta plain facies, but these facies also show the widest variation in coal thickness and are prone to channel cutoffs. In the Appalachians, the most successfully mined coals tend to occur in the transition between the lower and the upper delta plain. Similar relations appear to characterize the coal occurrences of the Natal Coal Field of South Africa (Hobday and Matthew, 1975).

Economic coal deposits within the Cenomanian–Santonian (Cretaceous) of Utah and other areas of the Rocky Mountains are associated with delta plain and deltaic depositional systems, but only where such systems are stacked vertically in association with minor transgressions and regressions superimposed on regional transgressive-regressive events (Ryer, 1983). Thick coals accumulated landward from paleoshorelines (Ryer, 1983). Where deltaic systems are associated with regional transgressive-regressive events, only thin coals occur. Coals associated with the Cretaceous Blackhawk Formation and Star Point Sandstone in Utah accumulated in both deltaic and barrier island settings (Flores et al., 1984). Coals associated with these deltas tend to be both thin and thick,

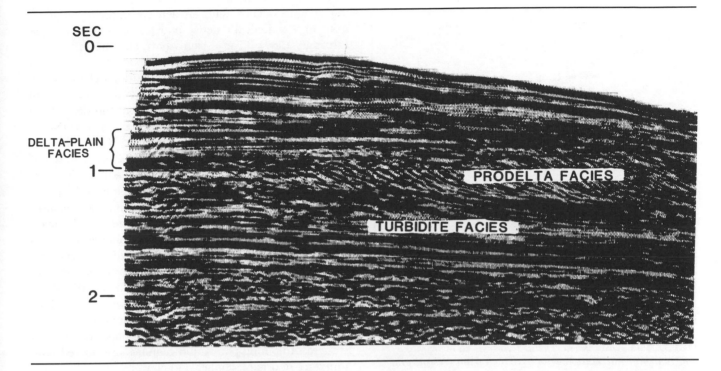

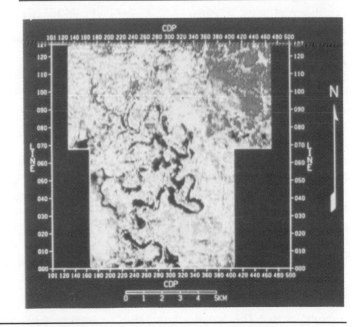

FIGURE 5.32. *Seismic section through river-dominated delta prograding over turbidites (seismic photograph courtesy of O.R. Berg; from O.R. Berg, 1982; republished with permission of the American Association of Petroleum Geologists).*

FIGURE 5.33. *Horizontal seismic section at depth of 196 msec showing meandering channel comprising part of Birdfoot Delta, the upper delta plain of which appears in uppermost center, Gulf of Thailand (Black and white photograph from Brown, Dahm, and Graebner, 1981; republished with permission of the European Society of Exploration Geophysicists).*

118

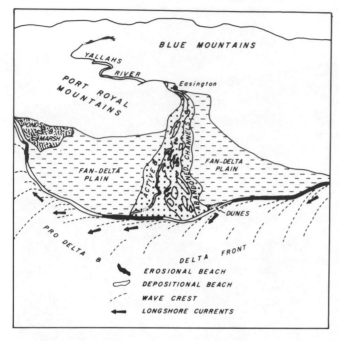

FIGURE 5.34. *Yallahs Fan Delta, Jamaica, showing major subenvironmental components (redrawn from Wescott and Ethridge, 1980).*

but discontinuous, whereas coals associated with barrier islands tend to be thicker and laterally continuous.

FAN DELTAS

In recent years, a different type of delta, common to active and passive continental margin settings, has been recognized. It is referred to as the fan delta, and it comprises a delta plain that is an alluvial fan and both delta front and prodelta zone (Wescott and Ethridge, 1980). Only two studies are published about Holocene fan deltas, the Yallahs Fan Delta in southeast Jamaica (Wescott and Ethridge, 1980) and the Copper River Delta of Alaska (Galloway, 1976). The Copper River Fan Delta is also influenced by glacial retreat.

HOLOCENE FAN DELTAS

Fan deltas represent the amalgamation of an alluvial fan substituting for a delta plain (in the sense of the Mississippi Delta), coupled with both the delta front and the prodelta zone. The Yallahs Fan Delta of southeast Ja-

maica is associated with an actively uplifted mountain front in a tectonically active margin zone (Wescott and Ethridge, 1980). River systems have cut into this mountain front (Blue Mountains) and debouche on a coastal plain (Figure 5.34). Because of the sudden change from a steep to shallow gradient, coupled with a change from confined to unconfined flow, an alluvial fan delta plain has formed. It is subdivided into active and abandoned channels, a fan-delta plain, local areas of marshes and ponds and, on the seaward edge, beaches and windblown dunes. The active channels are the main drainage water courses on the fan delta. On the floor of the channels, coarse gravel occurs mixed with sands and is fashioned into longitudinal bars. Both the sands and gravels show crude parallel stratification. Abandoned channels occur as a series of topographic lows filled with water for most of the year. Channel overbank deposition provides muddy sediment to these low areas and fills the abandoned channels. The remainder of the fan-delta plain is underlain by silt and sand organized into micro-cross-laminae, cut-and-fill structures, and roots and burrow features. Adjoining ponds and marshes are floored with a mixture of organic matter and clay.

Seaward from the Yallahs Fan is the delta front and prodelta zone. The delta edge consists of beach and dune sands and grades into a sloping zone composed of mixtures of sand, silt, clay, and gravel. The slope is relatively steep, and it is cut by canyons that are similar apparently to delta front troughs described by Shepard (1973). Slumping is known to occur in the heads of these canyons and, according to Wescott and Ethridge (1980), is the principal means by which sediment is transferred from the coastline and shoreface to deeper waters. Seaward from this zone is the prodelta zone, consisting of muds. Laterally, it was observed to grade into reefs.

The Copper River Fan Delta in Alaska (Galloway, 1976) shows many features that are similar to the Yallahs Fan Delta. What is different, however, is the greater marine energy flux influenced by a high mesotidal to macrotidal regime coupled with strong wave action. As a consequence, the tidal exchange is greater than the microtidal coast of the Yallahs Fan Delta, and barrier islands, tidal inlets, and tidal flats are welded onto the fan-delta edge. The inner part of the Copper River Fan Delta consists of distributary channels containing sand and gravel. This

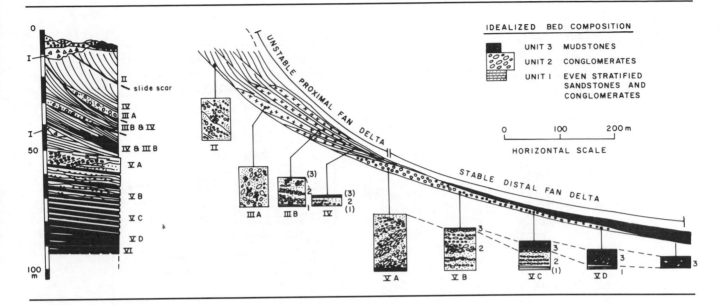

FIGURE 5.35. *Depositional model of fan delta in Abrioja Formation (Pliocene), Spain, showing major lithologic changes and associated fabric and structures (redrawn from Postma, 1984).*

delta plain zone grades seaward into a delta front or shoreface zone and prodelta zone consisting of clay.

ANCIENT FAN DELTAS

Several ancient fan deltas have been reported in the literature and include the Miocene–Pliocene Espiritu Santo Formation and Abrioja Formation of Spain (Postma, 1983, 1984), the Devonian of Norway (Pollard, Steel, and Undersrud, 1982), the Pennsylvanian of the Anadarko Basin (Dutton, 1982), the Eocene Wagwater Group of Jamaica (Wescott and Ethridge, 1983), and the Capo Blanco Sandstone of Peru (Palomino-Cardenas, 1976). The Abrioja Formation of Spain is perhaps the best-described ancient fan-delta sequence and it exhibits a complex association of sandstone, conglomerates, and mudstones (Postma, 1983, 1984). These contain a variety of structures including rippling, cross-stratification, and water-escape structure (Figure 5.35). Perhaps the most dominant sedimentary structure is the appearance of major slump scars (Figure 5.35), which brought fan-delta

plain sediments into the upper delta front zone (Postma, 1984).

FAN-DELTA VERTICAL SEQUENCE

Wescott and Ethridge (1980, 1983), Postma (1984), and Palomino-Cardenas (1976) suggested stratigraphic models and sections that indicate that as fan deltas prograded basinward, they tended to develop coarsening-upward sequences. The base of the sequence consists of prodelta muds, overlain by delta front mixed lithologies and slump deposits, capped by the fan-delta plain channel sands and gravels. A suggested sequence, based on field study from ancient examples (Postma, 1984; Palomino-Cardenas, 1976) and theorized from Holocene field work in the Yallahs Fan Delta (Wescott and Ethridge, 1980), is shown in Figure 5.36.

FAN-DELTA HYDROCARBON CASE HISTORIES

Hydrocarbon production from two fan-delta reservoirs is known. The first case is the Mobeetie Field in the Anadarko Basin of the Texas Panhandle (Dutton, 1982) and the second is from the Capo Blanco Sandstone (Eocene) of the Progresso, Talara, and Sechura Basins in Peru (Palomino-Cardenas, 1976). In the Mobeetie Field, braided-stream reservoirs on fan deltas are interbedded with car-

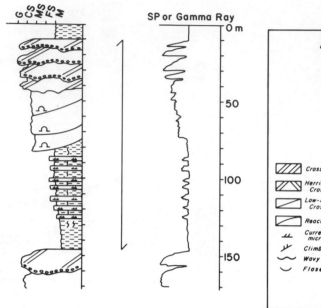

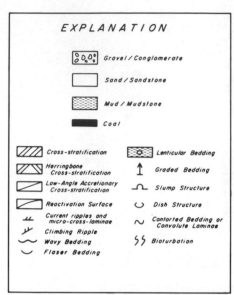

SP or Gamma Ray

EXPLANATION

	Gravel / Conglomerate
	Sand / Sandstone
	Mud / Mudstone
	Coal

	Cross-stratification		Lenticular Bedding
	Herringbone Cross-stratification		Graded Bedding
	Low-Angle Accretionary Cross-stratification		Slump Structure
	Reactivation Surface		Dish Structure
	Current ripples and micro-cross-laminae		Contorted Bedding or Convolute Laminae
	Climbing Ripple		Bioturbation
	Wavy Bedding		
	Flaser Bedding		

FIGURE 5.36. *Idealized vertical sequence and log motif, fan delta system. Doubly terminated arrow shows sequence interval. (Abbreviations: G—Gravel; CS—Coarse sand; MS—Medium sand; FS—Fine sand; M—Mud.)*

bonate reservoirs. The distal zone of the fan delta was reworked by marine processes, but is not of reservoir quality because of carbonate cementation. In the Capo Blanco Sandstone (Eocene), the principal reservoirs are in the braided-channel complex of the fan-delta zone. They comprise the upper part of a coarsening-upward sequence (Figure 5.37) and the log shapes show typical interdigitation of braided-channel systems. The Capo Blanco Sandstone fan-delta complex overlies delta front and delta plain facies of the Clavell Formation (Figure 5.37). It formed in response to rapid rates of uplift along the Andes during Eocene time.

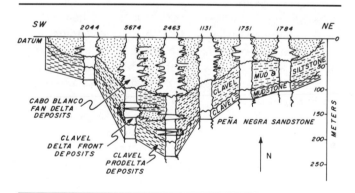

FIGURE 5.37. *Electric log and subsurface stratigraphic cross-section of Capo Blanco Sandstone, northwest Peru (redrawn from Palomino-Cardenas, 1976).*

Chapter 6

Continental Shelf Sand Bodies

INTRODUCTION

The continental shelf environment comprises a depositional zone characterized by moderate to low slope and ranging in depth from the lower shore face (about 5 m depth) to a major break in slope occurring at an average depth of approximately 200 m below sea level. This depositional zone occurs also between areas dominated by nearshore processes and those dominated by oceanic slope and deep-water processes. The shelf environment occurs in several types of continental margins (Emery, 1980) including both passive margins flanking deep ocean basins associated with active spreading centers, such as the continental shelves of the Atlantic Ocean, shelves associated with active margins possessing subduction zones and complex basins (such as along the circum-Pacific), and broad epicontinental seas possessing a basement of continental crust such as the Yellow Sea and the North Sea.

Although understanding of continental shelf sedimentation processes is relatively new, studies of continental shelves go back to the early part of this century. Johnson (1919) theorized from morphological data that continental shelves were characterized by a profile of equilibrium (Figure 6.1) and that their surface sediment distribution showed a progressive decrease in grain size from the shoreline to the shelf edge. This laterally graded sediment distribution was controlled by a concomitant decrease in bedshear and bottom-current velocity. Later, Shephard (1932) completed a sampling survey of the Atlantic continental shelf of the USA and demonstrated that its sediment distribution was characterized by a great deal more

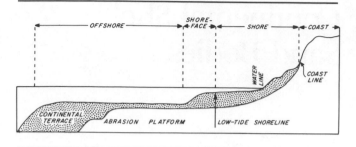

FIGURE 6.1. *Hypothetical profile of equilibrium of a continental shelf (redrawn from Johnson, 1919).*

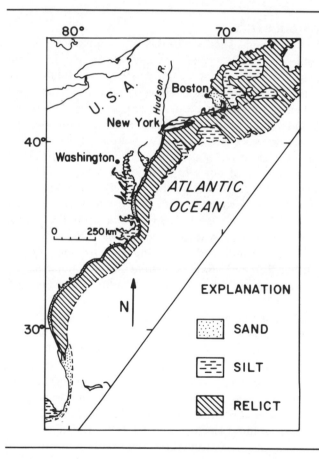

FIGURE 6.2. *Continental shelf of eastern USA showing distribution of interpreted relict areas of sedimentation and active areas of sand and silt deposition (redrawn from Emery, 1968).*

complexity than Johnson's (1919) hypothesis predicted, and gravels were not uncommon near the shelf edge. This sediment distribution was related in part to eustatic lower stands of sea level during the Pleistocene, which would expose much of the shelf surface and superimpose on it a variety of nonmarine and coastal processes expressed as remnants of peat deposits, channels, barrier islands, and coastal dunes (Milliman, Pilkey, and Ross, 1972; Emery et al., 1967; Field et al., 1979; Field and Duane, 1976). Because of a relatively rapid rise in sea level (Shepard, 1960; Emery, 1968; see also Chapter 4), these sediments were drowned and left largely undisturbed. These sediments were termed "relict" by Shepard (1932) and Emery (1968), who mapped the distribution of such relict sediments globally (Figure 6.2).

Sediment accumulation on shelves is controlled partly by other factors also. Emery (1968) demonstrated that some accumulation can occur behind several types of dams including large reefs, diapirs, and tectonic fault blocks. Climate and latitude affect shelf sedimentation also (Emery, 1968), and therefore equatorial latitudes contain a larger volume of carbonate shelf sediments and reefs, while polar latitudes contain a larger volume of ice-rafted gravelly sediments. Moreover, coastal upwelling could contribute large volumes of phosphatic sediment on shelves.

SEDIMENTATION PROCESSES

Shelf sedimentation processes are variable. They include tidal currents (Off, 1963; Van Veen, 1935; Stride, 1963, 1982; Boggs, 1974, among others), wave-generated currents (Lavelle, Swift, Gadd, Stubblefield, Case, Brashear, and Haff, 1978; Lavelle, Young, Swift, and Clarke, 1978; Swift et al., 1977; Swift, Parker, Lanfredi, Perillo, and Figge, 1978; Swift, Sears, Bohlke, and Hunt, 1978), and wind-driven, storm-generated currents (Foristall, 1974; Lavelle, Swift, Gadd, Stubblefield, Case, Brashear, and Haff, 1978; Lavelle, Young, Swift, and Clarke, 1978; Gadd, Lavelle, and Swift, 1978; Swift et al., 1977; Swift, Parker, Lanfredi, Perillo, and Figge, 1978; Swift, Freeland, and Young, 1979; Hayes, 1967, among others). These processes may be persistent, such as tidal currents, or they may operate on a periodic or seasonal basis, as is the case with wave- and storm-dominated systems. Based on this realization, continental shelves are now characterized in terms of wave- and storm-dominated shelf systems and tide-dominated shelf systems. These shelf types represent some sort of end members, and gradational changes in between are most common (see also Hayes, 1967). Recently it was recognized that storms play a significant role in tide-dominated regimes as well (Nelson et al., 1982; Nelson, 1982). In the discussion to follow, shelf sand-body types are reviewed on a regional, case-by-case basis first because the end-member types are in preliminary form.

Emergence of the end-member concept of wave-storm versus tide-dominated shelf systems has come in part from a well-known oceanographic paradigm that may need modification. It is the principle that increasing shelf width causes an increase in coastal tidal range and a concomitant increase in bottom tidal-current velocities and bedshear (Sverdrup, Johnson, and Fleming, 1942; Redfield, 1958; Klein, 1977a,b; Klein and Ryer, 1978; Cram, 1979). This well-known principle holds that as continental shelves become progressively embayed, or progressively wider, tidal-current velocities, bedshear, and coastal tidal range increase (see also p. 75). Recently, Cram (1979) and Klein and Ryer (1978) published curves showing this correlation (Figure 6.3), although Cram (1979) has demonstrated also that the slope of these curves tends to differ from place to place. She attributed this difference to local physiography, differing amplitudes of ocean tides at the shelf break, and angle of approach of cotidal range lines. Tide-dominated shelves of the world such as the North Sea, Yellow Sea, and Bering Sea are characterized by large shelf widths, whereas on narrower shelves, tidal currents play a subordinate role.

In the discussion that follows, continental shelves are treated regionally to emphasize their variation and gradational change from those that are narrow and wave dominated, to those that are wide and tide dominated. That discussion focuses only on shelves where extensive sand bodies are now accumulating; this approach provides a predictive baseline for finding ancient counterpart reservoir sandstones in the subsurface.

ATLANTIC SHELF SAND BODIES OF EASTERN NORTH AMERICA

Morphological surveys (Shepard, 1932; Emery, 1968; Jordan, 1962; Swift, 1970, 1974; Ludwick, 1970, 1974, 1981;

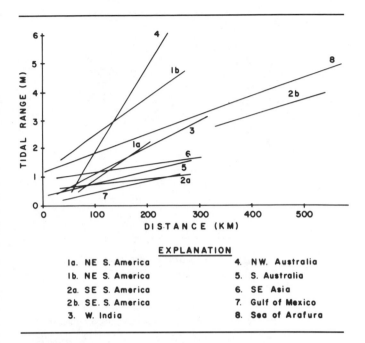

EXPLANATION

1a. NE S. America		4. NW. Australia	
1b. NE S. America		5. S. Australia	
2a. SE S. America		6. SE Asia	
2b. SE. S. America		7. Gulf of Mexico	
3. W. India		8. Sea of Arafura	

FIGURE 6.3. *Comparison of tidal range and distance to the 100-m isobath on continental shelves (redrawn from Cram, 1979 and Klein and Ryer, 1978).*

Swift et al., 1977; Swift, Parker, Lanfredi, Perillo, and Figge, 1978; Swift, Sears, Bohlke, and Hunt, 1978; Swift, Freeland, and Young, 1979; Field, 1980; Duane et al., 1972; Stubblefield, Kersey, and McGrail, 1983) demonstrated that large linear sand ridges are common to the Atlantic continental shelf of eastern North America and are associated with old channels and large shoals (Figure 6.4). Many of these are coast-parallel or coast-oblique with transitional relations tying the two together (Figure 6.5). Some coast-oblique ridges are welded onto the coastline also (Figure 6.5). Others are tidal-current sand ridges restricted to areas of constricted tidal flow or within larger embayments (Jordan, 1962; Smith, 1969; Ludwick, 1970, 1974, 1981). It is the coast-oblique and coast-parallel sand ridges that are of interest in this discussion.

One of the characteristics of the sand ridges in the Atlantic shelf of the eastern USA is that they are oriented parallel to each other over a large area (Figure 6.5). This morphological arrangement suggested to some that they

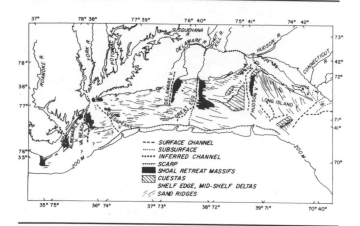

FIGURE 6.4. *Morphological features of the Middle Atlantic Bight, eastern USA (redrawn from Swift, 1974).*

are a series of relict barrier island chains that were subsequently reworked by present-day processes (Field and Duane, 1976; Stubblefield, Kersey, and McGrail, 1983). These features are preserved now as degraded barriers (Stubblefield, Kersey, and McGrail, 1983) that experienced substantial erosion of the barrier crestal zones and the seaward side.

Understanding of the origin of these features was advanced recently by observations of bottom-current velocities over seasonal changes (Gadd, Lavelle, and Swift, 1978; Lavelle, Swift, Gadd, Stubblefield, Case, Brashear, and Haff, 1978; Lavelle, Young, Swift, and Clarke, 1978; Swift and Field, 1981). These observations showed that during fair-weather periods, most sediment transport and bottom currents on the sand ridges were driven by wave-forcing or by relatively low-order velocity magnitude tidal currents (Lavelle, Swift, Gadd, Stubblefield, Case, Brashear, and Haff, 1978; Lavelle, Young, Swift, and Clarke, 1978). Sediment parameters appeared to be in equilibrium with these low-order velocity magnitudes, including grain size and bedform scale (Swift and Freeland, 1978; Swift, Freeland, and Young, 1979; Mann, Swift, and Perry, 1981). During storms, wind-forced currents combined actively with existing wave or tidal currents and bottom-current velocities were doubled (Lavelle, Swift, Gadd, Stubblefield, Case, Brashear, and Haff, 1978; La-

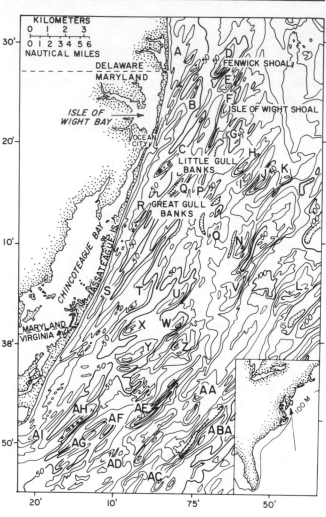

FIGURE 6.5. *Bathymetry of Assateague sand ridge field, Delmarva Peninsula, eastern USA. Depth contours in feet (redrawn from Swift and Field, 1981).*

velle, Young, Swift, and Clarke, 1978; Gadd, Lavelle, and Swift, 1978; Swift and Field, 1981). However, the sediment transport rate doubled also over a short duration (nearly two days) and accounted for 95% of the total sediment transport determined over a period of 135 days (Swift et al., 1981). A concomitant increase in suspended sediment rate was observed also. The flow direction of the storm-enhanced currents was parallel to the alignment of the sand ridges and a causal connection between them was inferred. In short, the origin of these sand bodies is a combination of perhaps a relict origin from eustatic low stands of sea level, shaping of coastal features (now drowned) during Holocene transgression, and redistribution, reshaping, and lateral extension parallel to flow directions by storm-generated currents, combined advectively with existing wave-generated and tidal bottom currents. These sand ridges represent a "palimpsest" association (Swift, Stanley, and Curray, 1971).

The ridges contain many features indicative of erosion and transport in a form not unlike a bedform. On the steepest side, erosional scours are common, as are coarser-grained sediments that are out of phase with the hydrodynamics, particularly on shorefaced ridges. Nearshore ridges and offshore ridges show features that are depositional and in equilibrium with a fair-weather bottom-current flow and later storm activity (Swift and Field, 1981). Fewer erosional features occur as these ridges are mapped in a seaward direction. Less modification occurs in deeper water because storm influence appears to diminish (Swift and Field, 1981). These ridge morphologies change from nearshore to midshelf because they represent a progressive evolution during transgression and later palimpsest modification and comprise part of a continuum with the land-connected ridges, representing the earliest stage, and the seaward ridges, representing the latest stages in equilibrium with present-day current and storm-enhanced current processes (Figure 6.6). Further modification of the ridges by combined wave-generated and storm-generated currents aids in causing a detachment of the sand bodies from original shoreline trends.

On the sand bodies themselves, the main transport direction of the reworked sediment is parallel to the sand ridges in the direction of prevailing storm-generated current systems. In the Atlantic shelf of the eastern USA, the

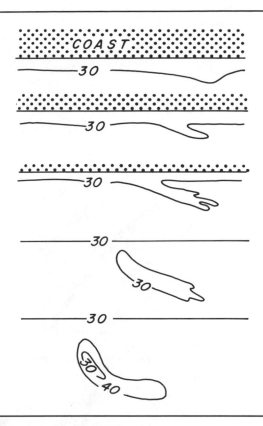

FIGURE 6.6. *Hypothetical sequence of growth and later detachment of shoreface sand ridge, Atlantic shelf, USA (redrawn from Swift and Field, 1981).*

prevailing wind directions flow from the northeast to the southwest, giving rise to the sand-body alignment observed on the seabed. Because the bulk of sediment transport occurs during these storms (Lavelle, Swift, Gadd, Stubblefield, Case, Brashear, and Haff, 1978; Lavelle, Young, Swift, and Clarke, 1978; Gadd et al., 1978; Swift and Field, 1981), sand dispersal patterns consist of unidirectional transport in the direction of storm-generated currents caused by prevailing winds during winter storms.

TIDAL-CURRENT SAND RIDGES, NORTH SEA
Linear sand ridges are common in the North Sea of western Europe. The earlier integrative work of Reineck

126

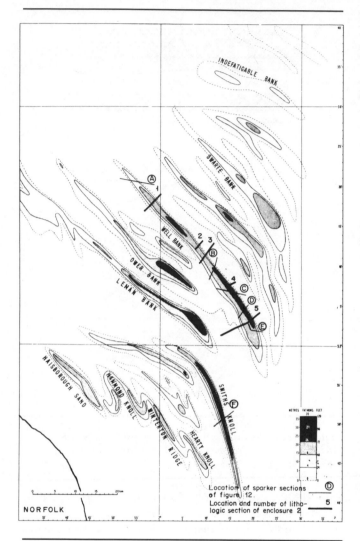

FIGURE 6.7. *Approximate isopachs of sand ridges, Well Bank, North Sea (from Houbolt, 1968; republished with permission of the Royal Netherlands Geological and Mining Society).*

(1963), Houbolt (1968), and Terwindt (1971) provided a baseline for understanding sediment dispersal and defining a vertical sequence. Other studies stressed the areal distribution of bedforms and sand-body trend with respect to flow directions of tidal currents on tide-dominated shelf seas (Stride, 1963, 1982; Caston, 1972; Jordan, 1962; Off, 1963; Belderson et al., 1972; McCave, 1970, 1971). Reineck (1963) was one of the earliest workers to describe the internal anatomy of the surface zones of such sand bodies, but his vertical sampling was limited to the upper 50 cm, a limit determined by his box-coring samplers. He demonstrated that the internal organization of the uppermost surface layers consisted of an extremely complex organization of cross-strata with thin sets (see also Houbolt, 1968; Terwindt, 1971; Stride, 1982).

Houbolt's (1968) study of the North Sea subtidal, tide-dominated sand bodies provided an insight concerning their internal structure because he combined sediment coring with continuous seismic profiling. These sand bodies are linear in plan and asymmetric in cross-section and are elongated parallel to both tidal-current flow directions and depositional strike (Figure 6.7). The surface is covered by dunes migrating toward the bar crest on the more gently sloping surface, and parallel to tidal flow on the steeper surface. Houbolt (1968) proposed from the bedform orientations that the sand bodies are subjected to a grain-circulation pattern through more gently sloping and more steeply sloping surfaces on these bars (Figure 6.8); this pattern is identical to the circulation pattern described by Klein (1970a) from the intertidal sand bodies of the Minas Basin of the Bay of Fundy. Seismic profiles (Figures 6.9 and 6.10) suggest an internal anatomy consisting of thick, steep cross-strata overlain by surface dunes and sand waves. The thick cross-strata dip in the same direction as the steeper slope of the sand body and presumably, the sand body migrates similar to dunes and sand waves. Such migration patterns are confirmed from morphological analysis of sinuous tidal sand bodies (Caston, 1972; Ludwick, 1974) and would generate a thick sheetlike sand body consisting of coalesced asymmetrical subtidal, tide-dominated sand bodies. These merged sand bodies would be stacked in an imbricate fashion separated by a clay layer, apparently generated by storm action (McCave, 1970, 1971). W.E. Evans (1970) has doc-

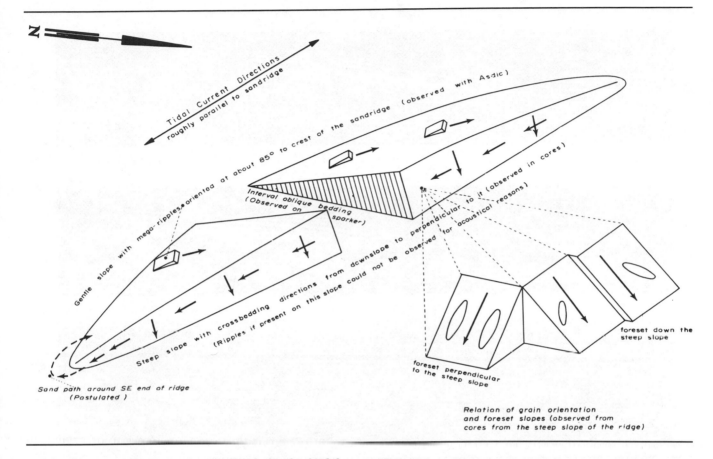

FIGURE 6.8. *Sediment dispersal model for idealized subtidal, tide-dominated sand body, North Sea (from Houbolt, 1968; republished with permission of the Royal Netherlands Geological and Mining Society).*

umented the preservation of such thin mud layers in a Cretaceous example.

It should be observed, however, that the seismic profiles of Houbolt (1968) in Figures 6.9 and 6.10 show considerable vertical exaggeration. Care must be taken then in interpreting the inclined reflections as thick cross-strata of the avalanche variety. Restoration of the profiles to their true angle indicates these cross-strata are in reality low-angle to nearly horizontal surfaces, or cross-planes. Presumably, they separate sets of thicker cross-strata, which thin upward. The cross-planes are thicker at the base

than at the top because they bound cross-strata formed by dunes that migrate in different water depths. Dunes migrating in greater water depths tend to be characterized by higher relief, generating thicker sets of cross-strata showing as thicker cross-planes in seismic section. Dunes generated in shallow water tend to be characterized by less relief, generating thinner cross-stratification sets. Such changes in bedform relief are observed on Houbolt's (1968) seismic records (see also Figure 6.10, Profile A). Consequently, the inner core of the tidal sand body would be characterized by thicker cross-strata and the surface zones by thinner cross-strata. Both core photographs of Reineck (1963) and Houbolt (1968), and Houbolt's (1968) seismic sections indicate these relations to be the case.

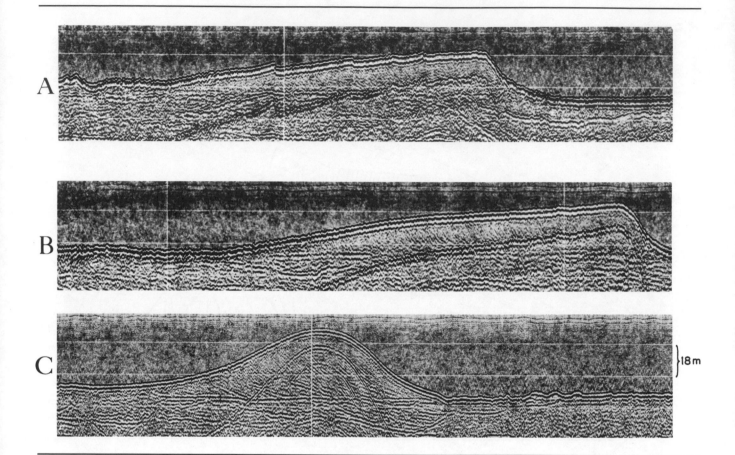

18 m

FIGURE 6.9. *Continuous seismic profiles over Well Bank and Smith Knoll, southern North Sea (from Houbolt, 1968; republished with permission of the Royal Netherlands Geological and Mining Society).*

The elliptical sand dispersal pattern that Houbolt (1968) determined from bedform alignment, internal reflectors, and regional tidal flow was challenged by later work in the North Sea. McCave (1979) demonstrated that bottom tidal currents in the North Sea at one location were rotary, rather than helical, and a different model would be required to account for the difference in current flow and bedform and morphological data. Later, Kenyon et al. (1981) suggested that sand transport over a tidal-current sand ridge involved both clockwise and anticlockwise transport paths in the direction of the bar slip face. Sand is dispersed parallel to the main sand body, then obliquely over the sand body in the direction of the slip face, and then parallel again on the opposite side of the sand body (Figure 6.11C). Kenyon et al. (1981) suggested

from regional current surveys of surface current flow and bathymetric maps that this mode of sand dispersal is characteristic of tidal current sand ridges in several examples, but later work by Klein et al. (1982), shown in Figure 6.11D, indicated a more complex origin of sand dispersal (discussed below). With the exception of McCave (1979), Klein (1970a), and Klein et al. (1982), none of the other dispersal models, such as proposed by Caston (1972, 1979), Stride (1963, 1982), or Belderson, Kenyon, and Stride (1971) and Belderson et al. (1972) are con-

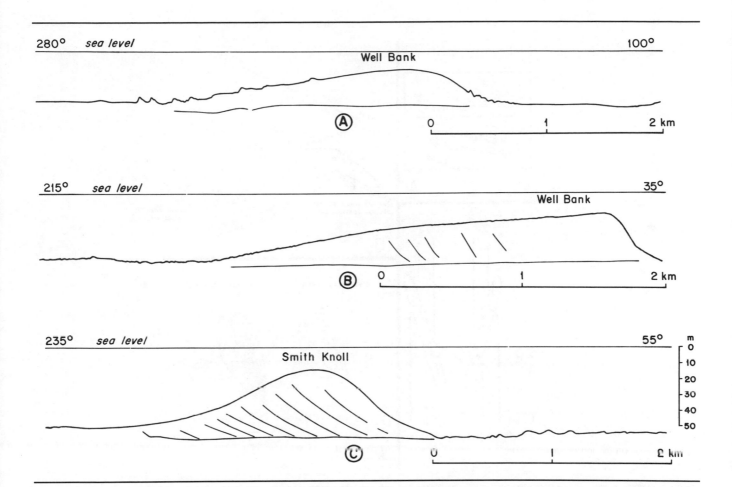

FIGURE 6.10. *Interpretation of continuous seismic profiles shown in Figure 6.9 (from Houbolt, 1968; republished with permission of the Royal Netherlands Geological and Mining Society).*

strained by obtaining data on bottom-current velocities and directions and transport rate.

One recent development with regard to tidal-current sand ridges in the North Sea is a better interpretation of the internal origin of compound sets and units of cross-stratification. Their organization appears to be controlled by the time–velocity asymmetry of tidal-current flow coupled with the longer-term alternation of neap and spring tides (Visser, 1980; Boersma and Terwindt, 1981; Nio, Siegenthaler, and Yang, 1983). It is well known that when maximum tidal-current velocities tend to be of nearly equal intensity, a vertical stacking of units of herringbone cross-stratification can be developed by accumulation of sediment coupled with opposite migration of dunes and sand waves with each reversing tidal phase (Reineck, 1963). The depth of sediment scour and reworking changes from the neap to the spring stage. During the change from the spring to the neap stage of a lunar tidal cycle, the depth of scour decreases, the velocity spectrum changes, and vertically stacked herringbone cross-stratification is preserved (Klein, 1970a). When time–velocity asymmetry is a dominant characteristic of tidal currents, reactivation surfaces (Klein, 1970) are formed during the subordinate-velocity phase and a unimodal orientation of

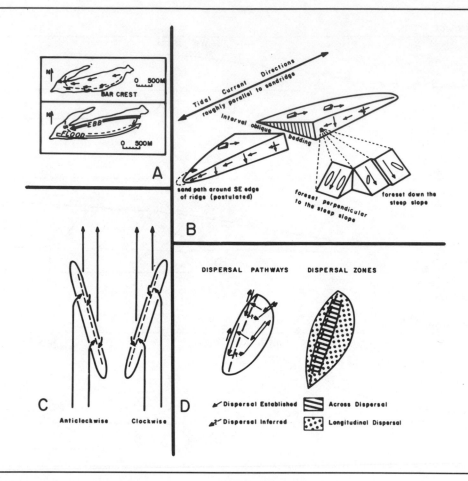

FIGURE 6.11. *Comparison of several modes of sand dispersal on tidal-current sand ridges. (A) Elliptical pattern, intertidal sand bar, Minas Basin, Bay of Fundy (after Klein, 1970a); (B) elliptical pattern, tidal-current sand ridge, North Sea (after Houbolt, 1968); (C) clockwise-anticlockwise pattern, several examples (after Kenyon et al., 1981); and (D) trapezoidal, tidal-current sand ridge, Yellow Sea, Korea (after Klein et al., 1982).*

cross-strata is observed, with bundles of such cross-strata bounded by reactivation surfaces dipping in the same direction, but truncating avalanche cross-stratification at a lower angle.

Superimposed on changes of flow directions of these tidal currents are longer-term cycles of magnitude changes of bottom-current velocities during the shift from neap to spring tide and back to neap tide. DeRaaf and Boersma (1971), Boothroyd and Hubbard (1975), Allen and Friend (1976), Visser (1980), and Boersma and Terwindt (1981) all observed that the migration distance of bedforms is greater during spring tides, and minimal, approaching zero, during neap tides. Boersma and Terwindt (1981) demonstrated that on one intertidal sand body, tidal-current velocities, characterized by time–velocity asymmetry, show larger mean and maximum bottom-current velocities and sediment transport rates during the spring phase of the lunar tidal cycle than during the neap phase (Figure 6.12). Within dunes and sand wave complexes in a flood-dominated area, where bottom-current velocity measurements were obtained, accretionary bundles of cross-strata with distinct bounding surfaces were observed within an individual set (Figure 6.13). Individual bundles of cross-strata were correlated to both neap and spring tide sediment transport. Reactivation surfaces (termed "pause planes") and associated structures graded laterally into a bundle of thick avalanche cross-strata organized into distinct laminae ("vortex structures" by Boersma and Terwindt, 1981). Laterally, these structures grade into a terminal interval of cross-strata possessing less well sorted sand with the angle of repose decreasing downcurrent. These are termed "slackening structures" (Figure 6.14). The entire sequence is overlain by sediments containing ebb-oriented cross-strata. This lateral change in structures records deposition during a single flood-dominated tidal phase, with the reactivation structure representing nondeposition during a subordinate phase, and the avalanche cross-strata representing the active phase of dune migration during the dominant flood phase. The slackening phase (Figure 6.14) represents the diminishing of flow velocities toward the end of a tidal cycle.

Internal organization of cross-strata and reactivation surfaces differs between the neap and spring phase because the neap phase shows thinner bundles and thinner cross-strata produced by smaller bottom-current velocities, whereas the spring tidal phase shows thicker sets of cross-strata and longer bundles, caused by greater velocities and greater sand transport rates. Thus the lateral dimensions and sediment volume permit recognition of spring and neap tidal phases in present-day tidal-current sand ridges. This interpretation can also be applied to ancient counterparts where neap pause planes may be represented instead by mud flaser beds (Allen, 1982; Nio, Siegenthaler, and Yang, 1983).

SAND BODIES ON THE BERING SEA SHELF

The epicontinental shelf of the Bering Sea differs from both the North Sea and the Atlantic continental shelf of

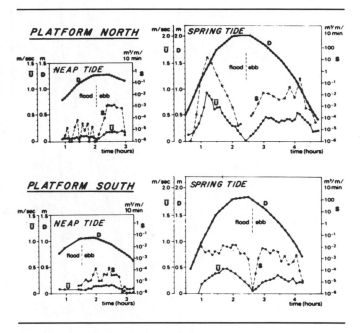

FIGURE 6.12. *Variation during a single tidal cycle of water depth (D) in meters, mean current velocity (U) in m/sec, and sand transport rate (S) in m³/m/10 minutes from two observation stations on intertidal sand body, Westerschelde, Netherlands, during spring and neap tide. Tidal-current velocities show time–velocity asymmetry as well as major change in velocity magnitude between spring and neap tide (redrawn from Boersma and Terwindt, 1981).*

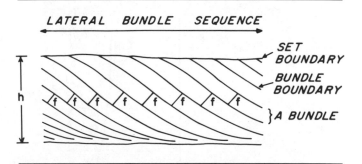

FIGURE 6.13. *Lateral bundle sequence and geometry within cross-stratified tidal-current sand ridge sediments. Bundle boundaries separated by mud or mudstone drapes. f = Thickness of successive bundles of cross-stratified sediment measured normal to bundle boundary (redrawn from Nio, Siegenthaler, and Yang, 1983).*

132

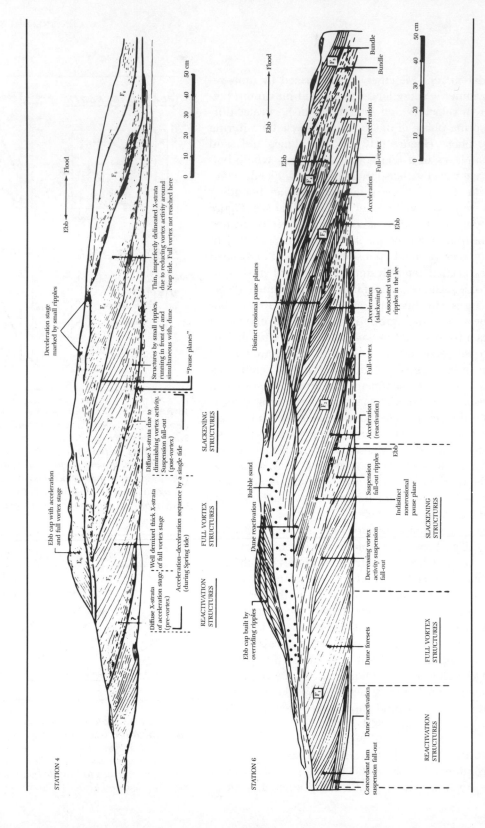

FIGURE 6.14. *Cross-section through intertidal sand body, Westerschelde Estuary, Netherlands, showing series of tidal bundles containing reactivation surfaces, slackening cross-stratification, and other evidence of change attributable to alternation of spring and neap tide (redrawn from Boersma and Terwindt, 1981).*

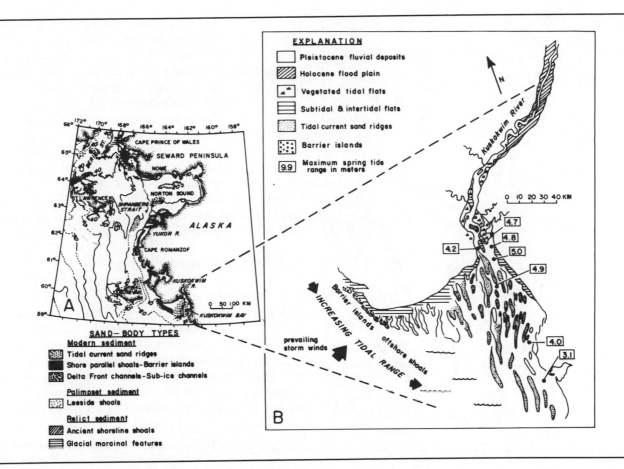

FIGURE 6.15. *(A) Variation in sand-body type in eastern Bering Sea shelf with (B) details of sand-body alignment and type and other sediments in Kuskokwim Bay, Alaska (redrawn from Nelson et al., 1982).*

the eastern USA because of its northerly latitude. As a consequence, its climatic regime is dominated by major cold air outbreaks during the winter months, and by riverine sediment yield and availability of glacially derived sediment (Nelson et al., 1982; Nelson, 1982; Drake et al., 1980). Sand-body types on the Bering shelf are extremely complex (Figure 6.15) and include tidal-current sand ridges, shore-parallel shoals and barrier islands, seaward extensions of delta front channel sands, relict sands, and palimpsest sand bodies.

Two areas are of interest because they were the sites of

detailed surveys: Norton Sound and Kuskokwim Bay. The sediment transport processes active in Norton Sound during fair-weather periods are dominated by tidal-current flow. Suspended sediment yield coming off the Yukon River Delta is relatively small (Drake et al., 1980). During stormy periods, suspended sediment concentration increases fivefold, and the current regime changes to combined wind-generated and oscillatory wave-generated currents (Drake et al., 1980). Transport of sand into Norton Sound from the Yukon Delta appears to be confined to periods of storms only (Drake et al., 1980).

Along Norton Sound, the coastline tends to be microtidal and mesotidal, and on its eastern side sedimentation is influenced strongly by sediment yield from the Yukon River (Nelson et al., 1982). Thus, a variety of bar features

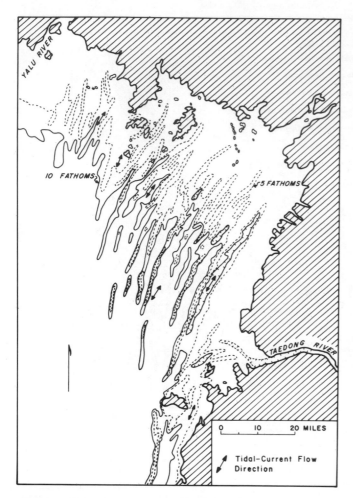

FIGURE 6.16. *Tidal-current sand ridges in the Gulf of Korea (redrawn from Off, 1963).*

are present, ranging from shore-parallel and barrier island systems with flanking tidal flats, to seaward extensions of distributary channels of the Yukon Delta (Figure 6.15A). These sand bodies range from 1 to 3 km long. During storms, sediment redistribution is common and sands are remobilized by cyclic wave loading (Nelson, 1982). This sand is not only graded, but is characterized by fining-upward Bouma sequences because these storm surge current systems are density currents also (Nelson, 1982).

Kuskokwim Bay differs only because it is characterized by a macrotidal coastline, probably because its embayed shape resonantly amplifies tidal range (Figure 6.15B). Here, tidal-current sand ridges are dominant and are oriented parallel to depositional strike on the east side of the bay, but at right angles to depositional strike at the bay head. These sand bodies exceed 10 km in length (Figure 6.15B). As one traces the sand bodies from the west side of Kuskokwim Bay to the bay head, tidal range increases from mesotidal to macrotidal and the sand-body type changes from coast-parallel barrier island to coast-normal tide-current sand ridges (Figure 6.15B). However, it is significant also that the bulk of sand transport occurs during storms, whereas at other times, modification of sand bodies is caused by tidal currents.

SAND BODIES IN THE YELLOW SEA

The Yellow Sea is the widest continental shelf in the world and is characterized both by a tide-dominated regime (Off, 1963; Niino and Emery, 1961; Chough, 1983, Asaoka and Moriyasu, 1966) and a macrotidal coast (Wells and Huh, 1980). The marine geology of the Yellow Sea was summarized first by Niino and Emery (1961) and more recently by Chough (1983). The sea floor shows an oceanward-sloping surface with semi-enclosed basins. Structurally, the Yellow Sea consists of a series of Neogene basin fills occurring behind tectonic dams of Precambrian and Mesozoic crystalline rocks and sediments (Emery et al., 1969; Wageman, Hilde, and Emery, 1970; Chough, 1983).

Water mass movements over the Yellow Sea are controlled seasonally (Niino and Emery, 1961; Asaoka and Moriyasu, 1966), with summer circulation dominated by the Kuroshio consisting of warmer saline waters, and winter currents dominated by colder, less saline waters, presumably altered by movement of southerly flowing high pressure cold air masses (cold air outbreaks) off the Siberian Platform (Wells, 1982). A strong tidal-current regime is superimposed on this circulation and accounts for the development of large tidal-current sand ridges (Off, 1963; Klein et al., 1982; Figure 6.16). As a consequence, the coastline is macrotidal with ranges from 4 to 9 m and more (Wells and Huh, 1980; Chough, 1983). Winter cold air outbreaks off the Siberian Platform cause a marked increase in suspended sediment load off the Korean coast; this material is transported off the shelf by a

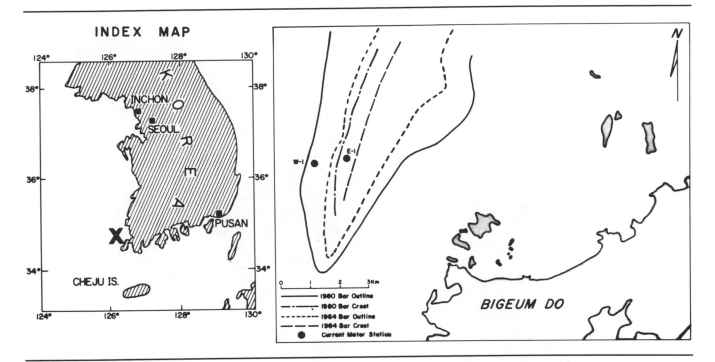

FIGURE 6.17. *Topographic changes from 1964 to 1980 on Odanam Satoe, a tidal-current sand ridge, Yellow Sea, Korea (redrawn from Klein et al., 1982).*

major storm-driven, coast-parallel boundary current off western Korea (Wells, 1982), which is accompanied by major set down of the water surface.

The sediment distribution of the Yellow Sea was summarized by Niino and Emery (1961), Park and Song (1971), and Chough (1983). The dominant sediment class is mud and most has been dispersed from both the Yangtze and Huang Ho rivers of China (Emery and Milliman, 1978). Sand in the Yellow Sea is modified into a series of tidal-current sand ridges (Off, 1963; Klein et al., 1982; Figure 6.16). Although Niino and Emery (1961) suggested that this sand was relict, Klein et al. (1982) documented both migration and accretionary growth of one sand body over a 16-yr period and from this finding, concluded that they were Holocene features (Figure 6.17). Accessory gravel deposits, probably of relict origin (Niino and Emery, 1961) also occur there.

Only one study (Klein et al., 1982) described bedforms

from the sea floor, and these were observed on and adjoining a single sand body (Figure 6.18). Plane bed surfaces, low-amplitude and high-amplitude dunes, and sand waves were identified by means of side-scan sonar mapping. The sand bodies are linear in plan and asymmetric in cross-section and a preferred orientation of bedforms was found in association with these changes in bar slopes. Thus on the steepest bar surface, plane beds were observed, whereas on the gentler bar surface, dunes were oriented with their crests parallel to the bar crest and dune slip faces were oriented in the same direction as the steepest bar slope. On the flanks of the sand body, dune slip faces are oriented parallel to the main flow direction of tidal currents, which is parallel also to the sand body axis.

Sand dispersal around the sand body was interpreted by Klein et al. (1982) to be trapezoidal because of the combination of time–velocity asymmetry of tidal currents on each side of the sand body, and observed rotary tidal cross-flow over the bar crest in the latest stages of ebb flow (Figure 6.11D). The source of these sands was suggested by Emery and Milliman (1978), Niino and Emery

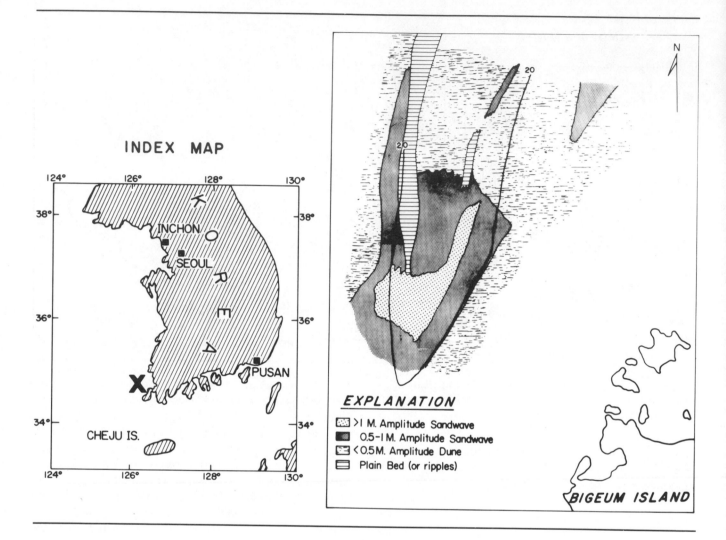

FIGURE 6.18. *Bedform distribution, Odanam Satoe, a tidal-current sand ridge, Yellow Sea, Korea (redrawn from Klein et al., 1982).*

(1961), Park and Song (1971), and Chough (1983) to be the Korean Peninsula, and they were thought to be dispersed into the Yellow Sea by local river systems such as the Han and Keum rivers. The sand bodies have been interpreted from bathymetry to be formed by tidal circulation (Off, 1963), a finding confirmed by Klein et al. (1982). It is not known what role winter storms and associated current systems play in the development and modification of these sand bodies, but both the coast-parallel alignment of these sand bodies in southwest Korea, and associated water set down on the surface of the Yel-

low Sea (Wells, 1982) provide a means for dispersing larger volumes of sand onto the sand bodies similar to the sand bodies in the epicontinental Bering Sea (Drake et al., 1980; Nelson et al., 1982).

SAND DISPERSAL ON CONTINENTAL SHELVES

The preceding discussion of four major continental shelf settings focused on sand-body origin and evolution. The nature of dispersal of sand in these shelves is variable. Two modes of dispersal in particular must be contrasted. First, sand dispersal from sources to their final sand bodies must be considered. Second, sand dispersal and transport on the individual sand bodies must be considered as a separate problem.

Dispersal of sand to continental shelves from potential sources has been mentioned in a few of the studies reported above. At least two potential sources exist. The first involves the distribution of sands over present-day shelf areas during eustatic lowering sea level by nonmarine and coastal processes. Some of these sand systems are preserved as shoal massifs such as in the Atlantic (Swift et al., 1977; Swift, Parker, Lanfredi, Perillo, and Figge, 1978; Swift, Sears, Bohlke, and Hunt, 1978), or relict bars such as in the Bering epicontinental shelf (Nelson et al., 1982). Moreover, many of these are being reworked by present-day storm-generated current systems, which cause alignment and accretion of the sand bodies in the downcurrent direction (Lavelle, Swift, Gadd, Stubblefield, Case, Brashear, and Haff, 1978; Lavelle, Young, Swift, and Clarke, 1978; Gadd, Lavelle, and Swift, 1978; Swift, 1981; Swift, Parker, Lanfredi, Perillo, and Figge, 1978; Swift, Sears, Bohlke, and Hunt, 1978; Swift and Field, 1981; Nelson et al., 1982), and are preserved as a palimpsest system.

The second source of sand is continental sediment yield associated with larger rivers. This type of sediment yield has given rise to the extension of distributary sand systems off deltas (Nelson et al., 1982) into the subtidal zone, but regional storm systems have caused resedimentation of this material and bar accretion also (Nelson et al., 1982) in the epicontinental Bering Sea shelf. Although not proven, the possibility exists that a similar process is causing accretionary growth of tidal-current sand ridges in the Yellow Sea (Klein et al., 1982). However, the sediment flux involved is not fully known.

On individual sand bodies themselves, the nature of sand dispersal during accumulation is unknown in the Atlantic shelf, but somewhat better known in tidal-current sand ridges (Figure 6.11). However, because the body of data used to determine these dispersal patterns differs, no preferred pattern has emerged. On intertidal sand bodies, Klein (1970a) reported an elliptical mode of dispersal alternating through flood- and ebb-dominated

time–velocity asymmetry zones (Figure 6.11A). His data were based on bedform mapping, tracer studies, and bottom-current velocity data. Houbolt (1968) inferred an identical dispersal pattern using bedform mapping with side-scan sonar, and mapping of internal reflectors on a tidal-current sand ridge in the North Sea (Figure 6.11B). Regional surface tidal currents were inferred, but not measured. Later, Kenyon et al. (1981) combined morphology, sand-body alignment, and surface tidal-current flow data to infer a linear flow parallel and across the tidal-current sand ridges in the direction of the steepest face of the sand bodies (Figure 6.11C). A more complex trapezoidal pattern (Figure 6.11D) was recognized in the Yellow Sea by Klein et al. (1982) from time–velocity asymmetry measurements on the bar flanks, rotary cross-flow of bottom tidal currents during the latest stages of the ebb phase of a tidal cycle, and from orientation of bedforms mapped with side-scan sonar. These patterns all share a fair-weather mode of transport and bar modification, presumably after storm reworking. However, the differences between the dispersal patterns may well depend on using direct measurements of bottom-current velocities and directions during a tidal cycle (Klein, 1970a; Klein et al., 1982), or just mapping surface features and combining that data with surface flow of tidal currents (Houbolt, 1968; Kenyon et al., 1981). As a consequence, the nature of these dispersal patterns still remains an open problem.

VERTICAL SEQUENCE OF CONTINENTAL SHELF SANDS

No overall synthesis of vertical sequences of continental shelf sand bodies exists at the present time. Most studies of continental shelves have focused on bar morphology, bedforms, and sediment dispersal.

Two possible vertical sequences are identified from a variety of data. Bouma et al. (1982) proposed a vertical sequence for wave-dominated and storm-dominated shelf sands comparable to what may be accumulating on the present-day Atlantic shelf (Figure 6.19). It is basically a coarsening-upward sequence consisting of a sharp base overlain by marine mudstones containing coquina layers towards the top. These graded upward into cross-bedded and micro-cross-laminated sands representing a marine bar facies. The marine sediments are comparable to the sand ridges described by Swift and his colleagues.

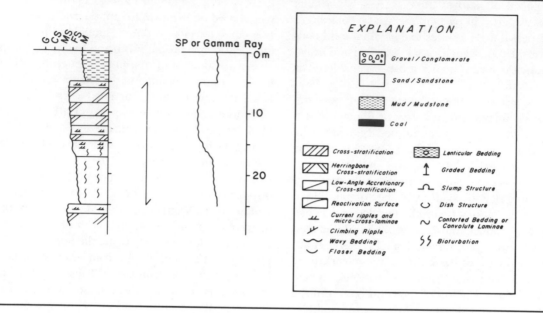

FIGURE 6.19. *Idealized vertical sequence and log motif, wave-dominated shelf. Doubly terminated arrow shows sequence interval. (Abbreviations: G—Gravel; CS—Coarse sand; MS—Medium sand; FS—Fine sand; M—Mud.)*

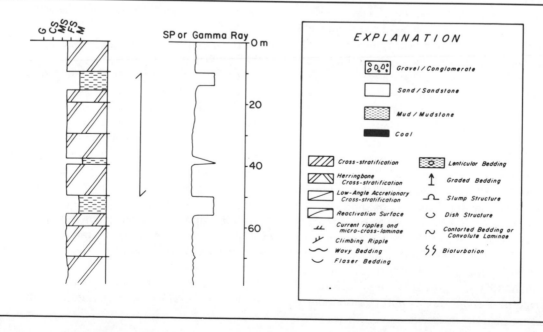

No specific facies vertical sequence model has been recognized from actual field mapping of Holocene, tidal-current sand ridges. However, two similar sequences were proposed by Klein (1977a) and by Johnson (1977); the former was developed by combining data published by Houbolt (1968), Reineck (1963), Caston (1972), Ludwick (1974), and W.E. Evans (1970), and the latter from outcrop study. Klein (1977a) proposed a hypothetical model (Figure 6.20) assuming the merging of multiple sand bodies into an imbricated sheet sand. Internally, a vertical sequence would consist of a basal core zone of relatively thick cross-stratified sand, overlain by an interval of thin cross-strata representing the upper reworked surface zone of active bedform migration. The top would consist of a thin clay representing the surface imbricated zone deposited during storm periods (McCave, 1970, 1971). The vertical sequence (Figure 6.20) consists of three parts: sand-body core (thicker cross-strata), the reworked surface zone (thinner cross-strata), and the clayey imbricated surface over which bars migrate to form a new sequence. Johnson's (1977) sequence differs from Klein's (1977a) only in that Johnson assumed the base of his sequence to be the clayey imbricated storm bed, with the sand body overlying that unit. In all other respects the sequences are identical.

More recently, Nio and Nelson (1982) reexamined this problem when comparing stratigraphic histories of the North Sea and the Bering Sea continental shelves. They argue that overall sequences should coarsen upward as Johnson (1977) and Klein (1977a) suggested, but the cause can be explained by accumulation of sediment during a rise in sea level. Thus, the basal muds may well be preserved tidal flats, and they are overlain by initial sand bank complexes, and these are capped by a sand wave sequence, representing the reworked surface zone by currents after accumulation of the sand banks reached an equilibrium (Figure 6.21). Similar sequences occur on open shelves, such as the seaward edges of the Bering, Yellow, or North Seas, or within shelf basins (Figure 6.21).

FIGURE 6.20. *Idealized vertical sequence and log motif, tidal-current sand ridge on tide-dominated shelf. Doubly terminated arrow shows sequence interval. (Abbreviations: G—Gravel; CS—Coarse sand; MS—Medium sand; FS—Fine sand; M—Mud.)*

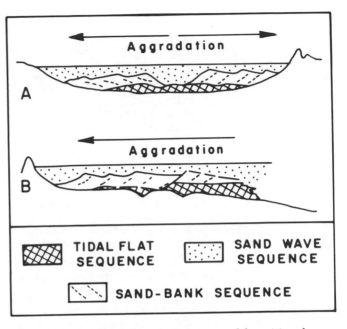

FIGURE 6.21. *Suggested vertical sequence of depositional systems in an epicontinental shelf basin (A) and an epicontinental shelf open to the sea (B) (redrawn from Nio and Nelson, 1982).*

ANCIENT EXAMPLES

As understanding of the nature of continental shelf sedimentation increased, a considerable number of sandstones were recognized as being formed under shelf conditions. A partial list of examples would include the Cape Sebastian Sandstone (Cretaceous) or Oregon (Bourgeois, 1980), the Cambrian Capito Formation of California (Mount, 1982), the Cardium Formation (Cretaceous) of Alberta (Walker, 1983a,b,c; Wright and Walker, 1981), part of the Silurian Tuscarora Formation of Pennsylvania (Cotter, 1983), the Upper Jurassic Stump and Swift formations of Wyoming and Montana (Brenner and Davies, 1973), the Martinsburg Formation of Virginia (Kreisa, 1981), the Sussex Sandstone (Cretaceous) of Wyoming (Brenner, 1978; R.R. Berg, 1975; Hobson, Fowler, and Beaumont, 1982), the Cretaceous Shannon Sandstone of Wyoming (Tillman and Martinsen, 1984), the Cretaceous Woodbine Sandstone of Texas (Turner and Conger, 1984), the Frio Sandstone of Texas (Berg and Powell, 1976), the Cretaceous sandstones within the Mancos

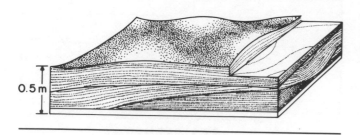

0.5 m

FIGURE 6.22. *Hummocky cross-stratification and associated hummocky surface (redrawn from Harms et al., 1975).*

Shale of Colorado (Boyles and Scott, 1982), the Permian Coolkilya Sandstone of Western Australia (Moore and Hocking, 1983), and the Miocene Molasse of Switzerland (Homewood and Allen, 1981). Most of these examples were considered counterparts primarily of the Atlantic shelf of eastern North America and were interpreted to have been deposited under a mixed-wave or accessory tidal regime with a major storm imprint.

Ancient counterparts of tide-dominated epicontinental shelf seas, such as the Yellow Sea of Korea or the North Sea, include the Precambrian Moodies Group of South Africa (Eriksson, 1977, 1979), the Precambrian Pongola Supergroup of South Africa (Von Brunn and Hobday, 1976), the Precambrian Pretoria Group of South Africa (Button and Vos, 1977), the Late Precambrian Lower Fine-grained Quartzite of Scotland (Klein, 1970b), the Precambrian and Cambrian Wood Canyon Formation of eastern California and Nevada (Klein, 1975a), the Cambrian Zabriskie Quartzite of eastern California and Nevada (Barnes and Klein, 1975), and the Ordovician Eureka Quartzite of the western USA (Klein, 1975b), among others. Perhaps one of the better-described examples of an ancient counterpart to a shallow, subtidal, tide-dominated sand body was described by Johnson (1977) from the Precambrian of Norway.

More recently, other examples of tide-dominated shelf sandstones have been reported by several workers. A partial list of examples includes the Permian Rancho Rojo Member of the Schnebly Hill Formation of Arizona (Blakey, 1982), the Jurassic Minette Oolitic Ironstones of Luxembourg (Teyssen, 1984), some of the Miocene Molasse of Switzerland (Nio, 1976; Allen and Homewood,

1984), the Precambrian Jura Quartzite of Scotland (Anderton, 1976), the Roda Sandstone of Spain (Nio, 1976), and the Lower Greensand and the Folkestone Beds of England (DeRaaf and Boersma, 1971; Nio, 1976; Allen, 1982). Most of these examples were identified as tide-dominated shelf counterparts from the presence of herringbone cross-strata, mudstone clay drapes over cross-strata, bioturbation features, and vertical sequences similar to those described by Johnson (1977), Klein (1977a,b), and Nio and Nelson (1982).

Counterparts to the present-day storm-dominated Atlantic shelf were recognized by a variety of features, of which hummocky stratification (Figure 6.22) is by far the most widely cited criterion. Hummocky stratification is characterized by lower bounding erosional set surfaces sloping at angles less than 10°, laminae overlying these surfaces in parallel arrangement, lateral thickening of laminae, and dip directions boxing the compass (Figure 6.22). Associated with the hummocky stratification are overlying flat laminae, micro-cross-laminae or cross-bedding, and bioturbated zones, which are expressed in that order as an idealized vertical sequence (Figure 6.23). Graded bedding may be associated with the hummocky intervals (Hamblin and Walker, 1979; Leckie and Walker, 1982; Mount, 1982; Dott and Bourgeois, 1982). These relationships suggest episodic sedimentation, particularly because the presence of bioturbated zones at the top of hummocky sequences (Dott and Bourgeois, 1982) indicates a period when conditions were favorable for colonization of the sea bed by burrowing organisms, before the next depositional event.

The process of formation of hummocky stratification is controversial. Most workers (see summary in Dott and Bourgeois, 1982) suggested development of hummocky stratification by orbital wave motion associated with storm processes. Nonetheless, as Dott and Bourgeois (1982) suggested, other possible processes may form the structure. Marsaglia and Klein (1983) suggested that breaking internal waves, separating stratified masses of opposite-flowing water, are characterized by large shearing stress and turbulence. Where such boundary conditions exist, these waves may impinge on the sea bed and provide a wavelike motion that could develop the hummocky beds. Moreover, Marsaglia and Klein (1983) demonstrated also that the paleogeographic position of many examples of hummocky stratigraphication occurs at a latitude too low to generate hurricanes, or on the side of continents where storm generation is rare. Thus, caution is advised in interpreting this structure.

More recently, Swift et al. (1983) reported a possible Holocene counterpart to ancient hummocky features in the Atlantic continental shelf. There, areas of dune bedforms are fairly common. However, during storm periods, side-scan data and diver's observations showed that these bedforms become rounded by scour, original spacing is preserved, but bedform relief is reduced. Internally these rounded bedforms contain subhorizontal laminae suggesting that they may contain hummocky stratification. Here the origin is attributed to a combined flow of unidirectional currents producing dunes, and later, short-lived superimposed, high-frequency, storm-generated wave orbital flow. These superimposed, storm-generated orbital currents would change unidirectional flow into a braided-flow pattern, altering dune morphology into a hummocky form. Bedform migration is suppressed under these conditions. Draping of suspended sediment from the storm load may account for the stratification features burying the rounded hummocky form. The role of the storm system is to superimpose an unstable flow pattern, which is combined with an existing unidirectional flow system. Preservation requires a rapid rate of sediment accumulation and burial from sediment placed in suspension by this unstable, superimposed flow. Thus, hummocky stratification is more likely to be characterized by a history of storm deposition, but this is not always the case.

CRATONIC DEPOSITION

Most of the ancient examples listed above are characteristic of epeiric and cratonic platform shelf seas, which share in common with continental shelves sediment deposition within relatively shallow water depths ranging from and including present-day shorelines to depths of water just below the photic zones. The scale of most cratonic depositional settings is comparable to broad and wide epicontinental shelf seas such as the North, Bering, and Yellow seas. These three modern shelf seas are clearly tide dominated because of the widely reported correlation between both increasing shelf width and increasing coastal tidal range (Figure 6.3) and increased

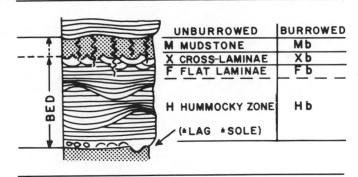

FIGURE 6.23. *Idealized vertical sequence of hummocky stratification and associated lithologies and structures (redrawn after Dott and Bourgeois, 1982).*

tidal-current bedshear on the sea bed. Thus, Klein (1977a,b) and Klein and Ryer (1978) argued that ancient cratonic systems ought to be tide dominated, and supported their argument by citing many Precambrian, Paleozoic, and Cretaceous examples.

More recently, Dott and Bourgeois (1982) cited many examples of cratonic sandstones containing well-developed hummocky stratification, and argued that perhaps ancient cratons were storm dominated rather than tide dominated. Other studies have indicated that eolian (Reiche, 1938; McKee, 1945; Wanless, 1981), fluvial (Potter and Pryor, 1961; R.R. Berg, 1980), and wave-dominated (R.R. Berg and Davies, 1968; Campbell, 1971; Sabins, 1962) sedimentation processes are represented also. Clearly, cratons of the past are characterized by a diversity of sedimentation processes.

Recently, Klein (1982) suggested that the variables controlling the timing of dominant sedimentation processes on cratons are a function of the lateral extent of marine inundation during a rise in sea level (thus modeling shelf width), tectonics in adjoining source terrains, and climate (controlling whether sandstones or carbonate systems occur). During the initial stages of a transgression, a craton is characterized by depositional processes comparable to the present Atlantic shelf of the eastern USA; the Cape Sebastian Sandstone (Cretaceous; Bourgeois, 1980) would be a typical counterpart. As sea level rises over a craton, the width of marine inundation (modeling shelf width) increases, and the regime changes to a tide-

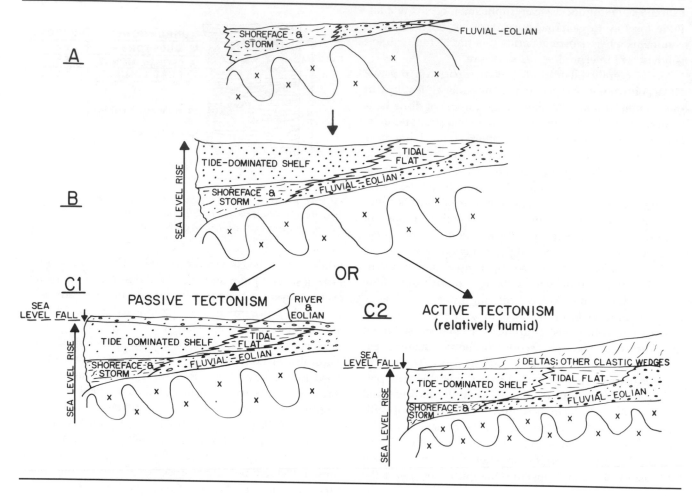

FIGURE 6.24. *Theoretical succession of depositional systems on a craton. (A) Early transgressive stage showing wave-dominated shelf sediments; (B) advanced stage of transgression showing later deposition of tide-dominated shelf systems tonguing laterally into tidal-flat systems; (C1) truncation of above succession with fall in sea level and regression of fluvial and eolian sediments; (C2) truncation of above succession, but instead deltaic clastic wedge is deposited on above sequence because of large rate of tectonic uplift in adjacent orogens (from Klein, 1982; republished with permission of the author and the Geological Society of America).*

dominated one comparable to the present-day Yellow Sea and North Sea. Should sea level fall, nonmarine fluvial and eolian systems are deposited, provided source area tectonics is inactive. Should source areas be characterized by a history of large tectonic uplift, deltaic depositional systems will accumulate. A synthetic stratigraphy was constructed to show these relationships (Figure 6.24). In seismic section, this stratigraphy would be expressed by a basal onlap sequence (Atlantic shelf type), overlain by a layered, concordant sequence (tide-dominated shelf counterpart), truncated either by fluvial-eolian beds, or by the sigmoidal signature of deltaic progradation. Several examples of these stratigraphic changes were re-

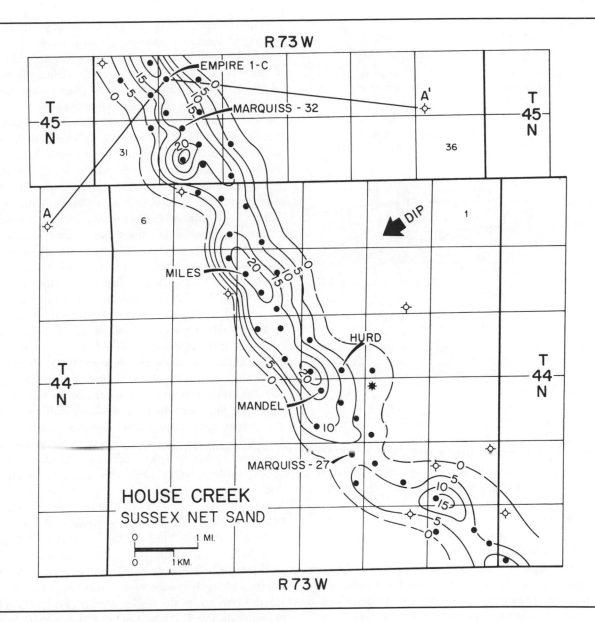

FIGURE 6.25. *Net sandstone isolith of Sussex Sandstone (Cretaceous), House Creek Field, WY. Contour interval is 5 ft (1.5 m). Isolith pattern suggests linear sand-body alignment parallel to depositional strike with an asymmetrical cross-section (from R.R. Berg, 1975; republished with permission of the American Association of Petroleum Geologists).*

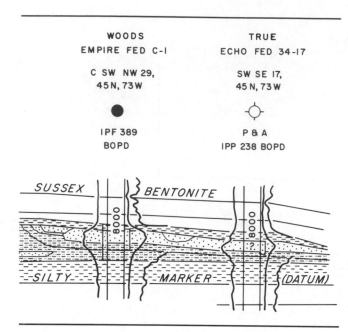

FIGURE 6.26. *Subsurface stratigraphic and log cross-section in downdip direction in Sussex Sandstone, House Creek Field, Wyoming (redrawn from Hobson, Fowler, and Beaumont, 1982).*

ported in support of the model. Thus, the counterpart Holocene shelf models discussed herein represent a gradational spectrum of the types of depositional styles that can occur on a craton; it is sea level fluctuations and source area tectonics that impose the principal controls on the lateral and vertical distribution of depositional systems on cratons.

OIL AND GAS CASE HISTORIES

Many of the ancient counterparts reviewed earlier are also petroleum reservoirs, particularly the examples reported from the Cretaceous of the Rocky Mountains such as the Sussex Sandstone. The Sussex Sandstone at House Creek Field, Powder River Basin, Wyoming, has been examined by several workers, most recently by Hobson, Fowler, and Beaumont, (1982), and earlier by R.R. Berg (1975) and Brenner (1978). They interpreted the Sussex Sandstone to represent a series of offshore bars deposited by storm-related and other marine currents, including both tidal and wave-generated currents.

The reservoir in question is linear in plan and asymmetric in section (Figure 6.25), and the sand bodies are aligned parallel to depositional strike. Examination of cores reveals a variety of sedimentary structures including shallow-water bioturbation, lenticular, flaser, and wavy bedding, channel floor lag concentrates, micro-cross-laminae, and cross-stratification (Hobson, Fowler, and Beaumont, 1982, their Figure 8; R.R. Berg, 1975, his Figure 4). Two types of electric log patterns have been reported. A funnel-shaped type pattern similar to that suggested for wave-dominated shelves (see also Figure 6.19) was reported by Hobson et al. (1982) within the lower part of the formation (Figure 6.26), whereas a blunt-base, blunt-top signature (Figure 6.27), more characteristic of tidal-current sand ridges (Figure 6.20), was reported by R.R. Berg (1975). Because it is not clear from either publication where these sand bodies occur with respect to sea level and older shoreline facies, it is not known whether the changes within the Sussex observed in the subsurface illustrate another example of the synthetic stratigraphy (Figure 6.24) suggested by Klein (1982). A seismic profile (Figure 6.28) obtained by Brenner (1978) shows the lateral extent of and the sharp acoustical impedance at both the top and the base of the sand body. Clearly, the core data, the log patterns, and seismic profile indicate that the Sussex Sandstone shares features in common with both Atlantic-type sand ridges such as those described by Swift and Field (1981) and tidal-current sand ridges (Off, 1963; Houbolt, 1968; Nelson et al., 1982; Klein et al., 1982).

R.R. Berg and Powell (1976) presented a second case of a sand ridge reservoir comparable to the present-day Atlantic shelf. It is from the Frio Sandstone (Oligocene), Nine-Mile Point Field, Texas, and shows evidence of a storm-deposited history. This evidence consists of shallow-water depth-graded beds similar to those reported by Hayes (1967) from the Gulf of Mexico, and Nelson (1982) from the Bering Sea. Reservoir sandstones include graded storm intervals as well as isolated sand bodies.

The Cretaceous Shannon Sandstone at Meadow Creek East Field, Shannon Field, Teapot Field, and Hartzog Draw, Wyoming, is another well-documented example of a petroleum reservoir that is a counterpart to the Atlantic shelf of eastern USA (Tillman and Martinsen, 1984). In both outcrop and subsurface, a complex set of facies are

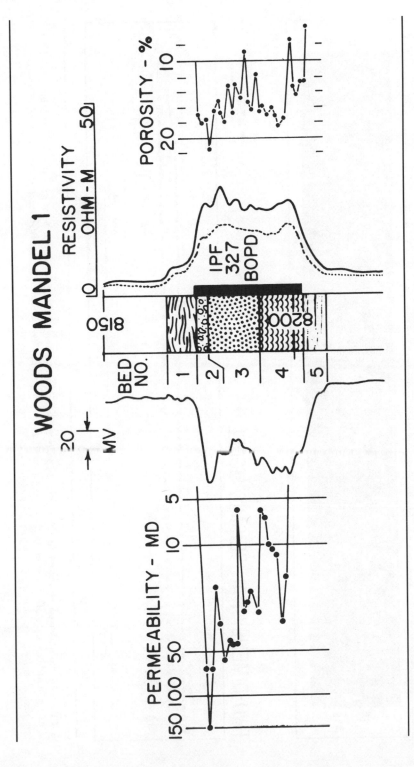

FIGURE 6.27. *Permeability and porosity of the Sussex Sandstone in Woods Petroleum Mandel Federal #1, House Creek Field, Wyoming, showing also blunt-base, blunt-top SP and resistivity log pattern (from R.R. Berg, 1975; republished with permission of the American Association of Petroleum Geologists).*

146

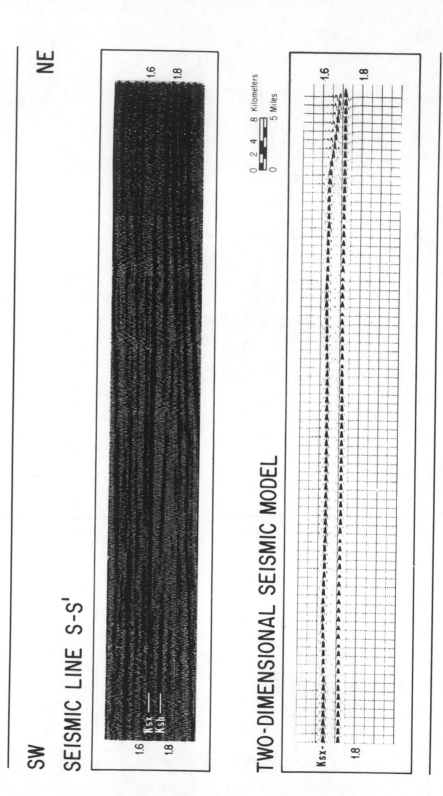

FIGURE 6.28. *Comparison of seismic line S–S' with two-dimensional seismic model showing eastern terminus of Sussex subtidal, tidal sand body (Ksx) and geometry of such sand bodies; Sussex Sandstone (Cretaceous) Powder River Basin, Wyoming (from Brenner, 1978; republished with permission of the American Association of Petroleum Geologists).*

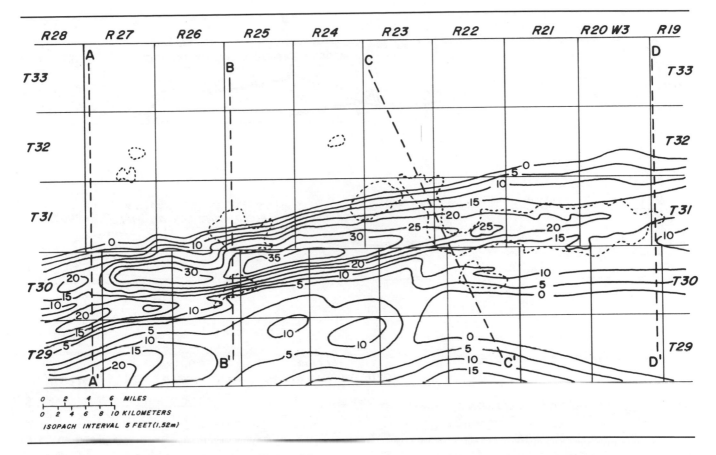

FIGURE 6.29. *Isopachous map of Cretaceous Lower "L" Member of Viking Formation, Doddsland Hoosier area, southwestern Saskatchewan. (Section line shown in Figure 6.30.) (Redrawn from W.E. Evans, 1970).*

present that all indicate a shelf environment with individual linear sand ridges preserved. Confirmation of this origin comes from vertical sequences in the surface and subsurface and electric log patterns that are similar to the wave-dominated shelf sequence and log patterns shown in Figure 6.19.

The Woodbine Sandstone (Cretaceous) at the Kurten Field, Texas, shows similarities to the sand bodies of the Atlantic shelf of the eastern USA (Turner and Conger, 1984). There sandstones are organized into a series of stacked linear sand bodies; each is isolated and linear in plan. The log shapes and vertical sequences again are

similar to the wave-dominated shelf sequences shown in Figure 6.19. Production appears to be confined to the top of the counterpart sand ridges, where reworking by currents presumably caused development of the best-sorted sands.

Two examples are suggested where petroleum and gas reservoirs show a tidal-current sand ridge history. The first case is the Cretaceous Viking Formation of southwestern Saskatchewan, Canada, where W.E. Evans (1970) documented sandstone reservoirs to be linear in plan and asymmetrical in cross-sections from isopach mapping (Figure 6.29), with an internal imbricate organization (Figure 6.30). These sand bodies were aligned parallel to depositional strike, thus confirming the interpretation.

Selley (1976), in a general study of electric log patterns and stratigraphic traps (Figure 1.1), provided a few case

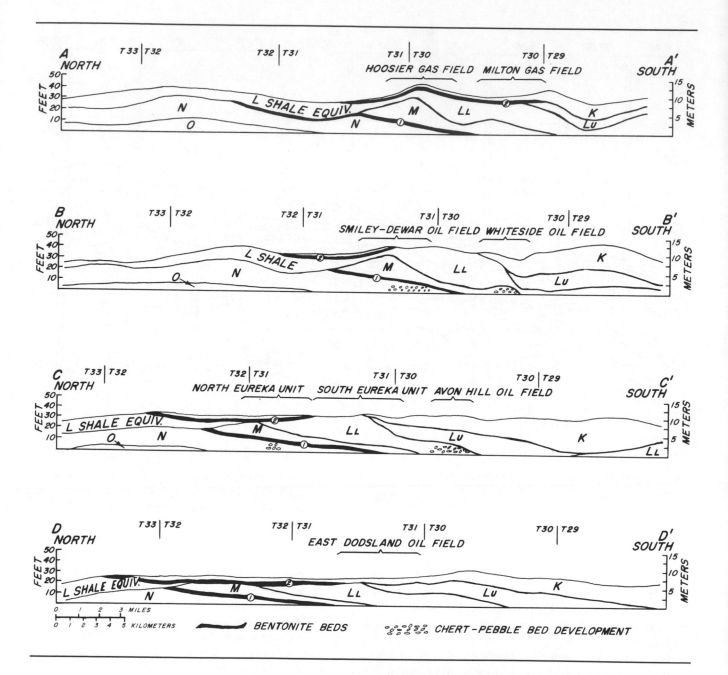

FIGURE 6.30. *Imbricate stacking of separate sandstone members of Viking Formation (Cretaceous), Doddsland Hoosier area, southwestern Saskatchewan. Sand bodies are asymmetric in cross-section separated by imbricate bentonite clays (redrawn from W.E. Evans, 1970).*

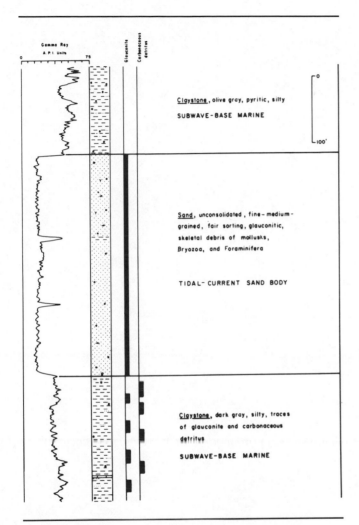

FIGURE 6.31. *Environmental analysis of part of North Sea well log following outline of Selley (1976). Tidal sand body shown in center (from Selley, 1976; republished with permission of the American Association of Petroleum Geologists).*

histories. One of these was from an undisclosed location and depth in the North Sea where a subtidal, tide-dominated sand-body origin was postulated (Figure 6.31). The characteristic SP and gamma-ray log pattern of such sands is blunt-base, blunt-top (Selley, 1976), and associated constituents include glauconite only.

Chapter 7

Turbidite Sandstone Bodies

INTRODUCTION

Turbidite sand bodies comprise part of a broad spectrum of sediments deposited by subaqueous gravity processes in deep-water marine environments. These processes are diverse in type (Figure 7.1) but all constitute part of a continuum. This continuum was classified formally by Middleton and Hampton (1973), who recognized four classes ranging from debris flow to grain flow to fluidized sediment flow to turbidity currents. Figure 7.1 shows a modification of their classification scheme and adds one other end member, namely the large submarine slumps and slides that have been described and recognized from several places (Moore, 1977; Booth, 1979; Doyle, Pilkey, and Woo, 1979; Keller, Lambert, and Bennett, 1979; Nardin et al., 1979; Schlee, Dillon, and Grow, 1979; among others). The sediment support mechanisms for each of these transport processes are different and permit segregation into distinct classes, as suggested by Middleton and Hampton (1973), although it must be emphasized that the roles of internal pore-pressure and liquefaction, as well as relative water content, make it difficult under natural conditions to observe these class distinctions so clearly (Lowe, 1976; Lawson, 1979a, 1982). Because water content, cohesion loss, and material strength changes interdependently (Figure 7.2), care must be taken in applying the classes recognized by Middleton and Hampton (1973) in too rigid a manner. This chapter reviews slump processes, as well as the four processes proposed by Middleton and Hampton (1973), and uses their terms to convey concepts in an easier and more organized fashion. However, the distinctions Lawson

SEDIMENT GRAVITY PROCESSES

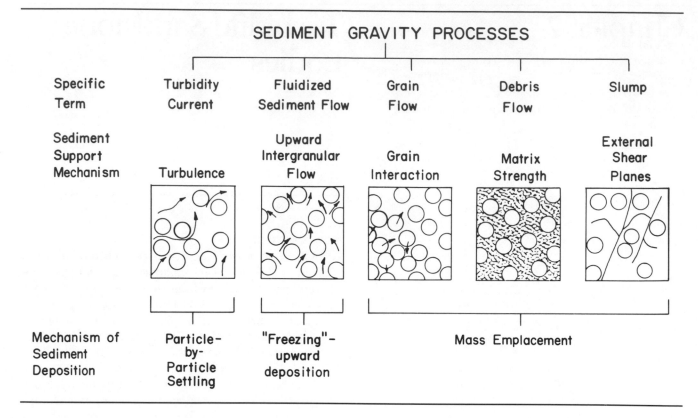

Specific Term	Turbidity Current	Fluidized Sediment Flow	Grain Flow	Debris Flow	Slump
Sediment Support Mechanism	Turbulence	Upward Intergranular Flow	Grain Interaction	Matrix Strength	External Shear Planes
Mechanism of Sediment Deposition	Particle-by-Particle Settling	"Freezing"-upward deposition	Mass Emplacement		

FIGURE 7.1. *Revised classification of subaqueous gravity processes, sediment support mechanisms, and mechanisms of deposition (redrawn partly after Middleton and Hampton, 1973).*

(1982) made in a subaerial counterpart (Figure 7.2) portray more realistically what one observes in nature.

This chapter focuses on subaqueous gravity deposits common to deep-water marine environments. However, it must be emphasized that their occurrence is not confined exclusively to the floors of ocean basins, continental margins, or submarine canyons, and submarine fans. Submarine slumps, debris flows, mudflows, and turbidity currents are common to deltaic settings (see Chapter 5 for review). Turbidity currents occur as hyperpycnal jet inflow off the Mississippi Delta during spring flooding (Scruton, 1956) and in Lake Geneva off the Rhone Delta (Houbolt and Jonkers, 1968). Large-scale slumps were reported off the Magdalena Delta by Shepard (1973), and a potential ancient counterpart of such depositional processes was documented from a Cretaceous deltaic system in the Reconcavo Basin of Brazil (Klein, DeMelo, and Della Favera, 1972). More recently, Lawson (1982) doc-

umented a broad range of nearly identical subaerial gravity processes and sediments from the edge of the Matanuska Glacier in Alaska (Figure 7.2). Thus the discussion of gravity processes in this chapter is also relevant to subaerial deposits (glacier margin; alluvial fan—see Chapter 2) and deltas, as well as deep-water settings.

GRAVITY PROCESSES

SLUMPS AND SLIDES

Slumps and slides are masses of sediments that have moved downslope along external shear planes, yet retained their internal coherence. These masses of sediment are variable in size, ranging in excess of 900 km³ by volume and covering an areal extent in excess of 2,000 km (Moore, 1977). The coherent mass of sediment may retain original stratification and other features, or may show evidence of deformation in the form of folds or internal fractures. Most of the fractures are extensional in origin, whereas most of the slump folds appear to be related to differential drag of coherent sediments adjoining the external shear plane of failure along which the slumped mass moves, or compressional stresses. The support mechanism for slumps and slides is a combination of retention of internal coherence of the sediment mass as well as the external shear planes of failure along which the slumps or slide masses move. The triggering mechanism for slump appears to be failure owing to shock such as earthquakes (Moore, 1977; Heezen and Ewing, 1952), fluctuations in Pleistocene sea level (McGregor, 1977), excess pore-pressure in greatly water-saturated, fine-grained sediments that experienced rapid burial (Booth, 1979), and oversteepened slopes (Keller, Lambert, and Bennet, 1979; Doyle, Pilkey, and Woo, 1979) developed either by rapid deposition or by localized tectonic folding on continental slope settings of active margins (Hampton and Bouma, 1977; Carlson and Molnia, 1977).

Sediment deposition is by mass emplacement. Such mass emplacement may be aided by reduction in slope, loss of fluid pore-pressure along the planes of failure, increased frictional drag along the planes of failure, or loss of coherence of sediment by progressive liquefaction and fluidization.

Recognition of the importance of slide and slumps in deep-water marine sediments was aided by the development of air gun seismic profiling. Although slumps were identified in association with cable breaks (Heezen and Ewing, 1952), it was not until the 1960s and 1970s that it was possible to determine the areal extent of these masses (Moore, 1977). Excellent examples of seismic profiles showing such slump features were published in recent years by Moore (1977), McGregor (1977), Carlson and Molnia (1977), and Schlee, Dillon, and Grow (1979), among others. The reader is referred to these papers for more details. One inference that can be made about these large slides reported from deep-water marine environments is that they are extremely common and show a large preservation potential. They should be preserved, therefore, in the rock record. Small-scale slump features

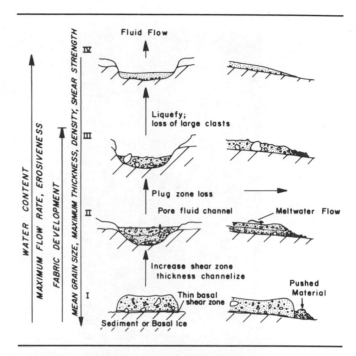

FIGURE 7.2. *Examples of four transitional types of sediment flows shown in transverse and downflow cross-sections from the terminus zone of the Matanuska Glacier. These sediment flows represent a continuum with changing water content, fabric, density, and shear strength (redrawn from Lawson, 1982).*

can be identified in outcrop, provided that the master bedding above and below the disturbed zone is undisturbed and parallel. However, because of the seismic profiles shown in Moore (1977), McGregor (1977), and Hampton and Bouma (1977), among others, one wonders how many so-called major fault systems that have been mapped in the past may, in fact, represent large-scale planes of failure along which slumping occurred. This problem still awaits solution.

DEBRIS FLOW

This discussion follows material reviewed by Hampton (1972, 1979) and Middleton and Hampton (1973). Debris flows are sediment–water mixtures that move along internal shear planes within the body of the material. These shear planes provide the means for sedimentary material to move downslope. Some of this mass moves as a rigid

FIGURE 7.3. *Debris-flow conglomerate showing dispersed clast fabric of volcanic and limestone clast set in matrix of nannoplankton limestone, Site 313, Core 27, Section 1 (82–118 cm), northern edge, Mid-Pacific Mountains, Pacific Ocean (20°10.53′N; 170°57.15′W).*

plug bounded by a zone along which shearing and movement occurs. Deformation of material is internal, and it includes loss of cohesion and remobilization of sediment within the zone of shearing and movement. Debris flow refers to sluggish movement downslope of mixed granular solids, clay minerals, and water in response to gravity (Hampton, 1972).

The granular solids, which are mostly sand and gravel, in a debris flow are transported and supported by the interstitial mixtures of clay and water. This process develops a supporting mechanism of *matrix strength* provided by clay minerals and water combined in a fluid style (Hampton, 1972; Middleton and Hampton, 1973, 1976). This matrix material is characterized by a limited cohesive strength that supports the floating grains of sand and gravel. The clay–water fluid shows an excessive density with respect to normal water, giving the debris flow greater buoyancy (Hampton, 1979) to support sand and gravel. Excess pore-pressure and large sediment concentration can aid this buoyancy and provide a more mobile flow (Hampton, 1979). Competency of the debris flow is a function of the strength and density of the clay mixture. Consequently, the greater the density, the coarser the granular solids that are transported. Once density diminishes, so does the largest grain size that the debris is able to transport (Hampton, 1972).

The solids are transported by a process of dispersion by which each of the grains exerts a dispersive stress provided by continual grain collision. The resulting sediments are of mixed grain sizes, ranging from very fine clay to extremely coarse gravel clasts. A fabric of coarser grains floating in finer grains results. This dispersed-clast fabric (Figure 7.3) is diagnostic of debris-flow sediments. Stratification is nonexistent.

Deposition occurs by a process of mass emplacement. The resulting textures appear like pebbly mudstones or tillites (Figure 7.3). They can be distinguished from tillites because debris flows show a random fabric (Lawson,

1979b; Taira and Scholle, 1979). Such deposits are common to a variety of deep-water settings and have been observed in cores obtained by the Deep Sea Drilling Project (Bouma and Pluenneke, 1975; Klein, 1975c; White et al., 1980), generally associated with inner fan deposits in back-arc basins and trenches or off submarine volcanoes (Cook, Jenkyns, and Kelts, 1976). In Holocene deep-water marine sediments, debris flows are also known to be spread over large areas as much as 800 km from the source (Embley, 1976; Kidd and Roberts, 1982).

The vertical sequence produced by debris-flow deposits is not understood clearly. A suggested sequence proposed by Middleton and Hampton (1973) is shown in Figure 7.4.

GRAIN FLOW

The discussion that follows is condensed from Middleton and Hampton (1973). The concept of grain flow was proposed by Bagnold (1954) to indicate a mechanism of sediment transport whereby upward supporting stresses act on grains within flowing sediments because of grain-to-grain collisions. The stress is grain-dispersive stress and it is proportional to the shear stress transmitted between grains. It prevents grains from being deposited out of flows. Sand avalanching down the slip face of an eolian dune where the pull of gravity moves the grains downslope represents a typical grain flow. Support is provided by grain interaction.

These kinds of flows have been reported by Dill (1964) from the upper ends of submarine canyons and were termed "rivers of sand." Such sand flows are strong enough to erode submarine canyons and the dispersive stresses are large enough to support gravel.

The depositional mechanism is, again, mass emplacement. The boundary of sedimentary deposits so emplaced is sharp and deposits are thick. The fabric consists of dispersed clasts of pebbles floating in sand, and the sediment texture ranges from clay to gravel (Klein, DeMelo, and Della Favera, 1972). The base of the beds may include sole marks, slide marks, and load structures, and internally may include dish structures. Stauffer (1967) interpreted the dish structures to be produced by grain flow, although more recent work has demonstrated that they form by water escape (Lowe and LoPiccolo, 1974), or by remolding of unstable portions of debris flows after de-

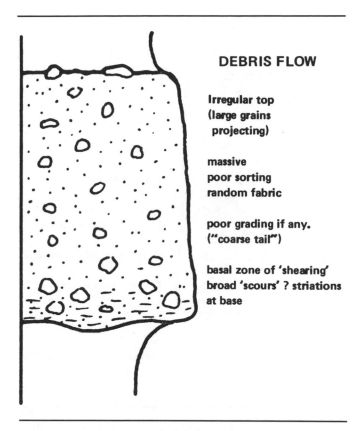

DEBRIS FLOW

Irregular top (large grains projecting)

massive poor sorting random fabric

poor grading if any. ("coarse tail")

basal zone of 'shearing' broad 'scours' ? striations at base

FIGURE 7.4. *Hypothetical vertical sequence of texture, sedimentary structures, and surface contacts in debris flow (redrawn after Middleton and Hampton, 1973).*

bris-flow deposition (Busch, 1976). Thus, it is noteworthy that perhaps the criteria suggested for some of the grain-flow features may indeed represent a different process of deposition. It is therefore quite likely that true grain-flow deposits may be rare in deep-water settings and are characterized by a poor preservation potential.

FLUIDIZED SEDIMENT FLOW

This type of gravity flow involves the expansion of a sediment bed by the introduction of fluids within the pore spaces and the fluid flow is such that both the fluid and the grains are moved upward. The sediments, in other words, become liquefied (Middleton and Hampton, 1973; Lowe, 1976; Lawson, 1982). The sediments must be

156

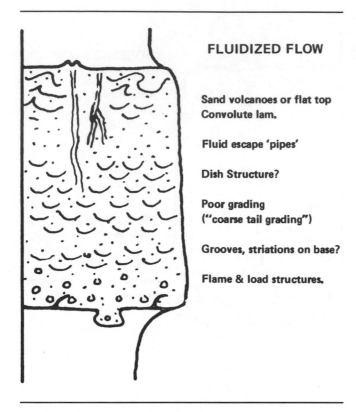

FLUIDIZED FLOW

**Sand volcanoes or flat top
Convolute lam.**

Fluid escape 'pipes'

Dish Structure?

**Poor grading
("coarse tail grading")**

Grooves, striations on base?

Flame & load structures.

FIGURE 7.5. *Hypothetical vertical sequence of texture, sedimentary structures, and surface contacts in fluidized sediment flow (redrawn after Middleton and Hampton, 1973).*

packed loosely for them to overcome the resistance of grain fabric to fluid injection. The supporting mechanism for this process is the excess pore-pressure of interstitial fluid that keeps the sedimentary particles afloat. Gravity propels the fluidized sediment downslope. A large concentration of sediment with excess pore-pressure is required to induce this flow mechanism.

To maintain the fluidized sediment in transport, pore-pressures must exceed hydrostatic pressures; the resulting sediment–fluid mixture would show little strength. The pore-pressure dissipates rapidly, and as it does so, deposition of sediment occurs. Deposition takes place rapidly from the base upward; this mechanism was referred to as a gradual "freezing-upward" mechanism (Middleton and Hampton, 1973, p. 14) from the base upward.

The sediment particle sizes transported by this process are clay, silt, and sand. Sediment concentrations are large, so only partial sorting and grading occurs (coarse-tail grading). Planar lamination, lensoid lamination, and lobate external forms are common (Lawson, 1979a). Dish structures may occur owing to pore-pressure loss and associated water escape, as suggested by Lowe and Lo-Piccolo (1974) and Middleton and Hampton (1973). Water escape pipes and sand volcanoes may also be present. A suggested vertical sequence of fluidized sediment flows appears in Figure 7.5.

RELATIONSHIP BETWEEN THESE FLOW PROCESSES

As indicated earlier, the above terms follow a classification suggested by Middleton and Hampton (1973, 1976) to distinguish between different processes of gravity sediment transport. It must be emphasized again, however, that these terms and classes are part of a continuum (Lawson, 1982); and furthermore, one type of flow may or *may not* be a precursor for one of the other classes. Thus, it is conceivable that debris flows can, with additional admixing of water, move as slurries or fluidized sediment flows. The reverse could happen as a result of rapid water loss and remolding. The classification Middleton and Hampton (1973) developed is useful to contrast different depositional processes and their support mechanisms, but no transition from one to the other, or succession of processes, is implied.

Some of these processes can occur rapidly. Thus, an undisturbed sediment can liquefy rapidly (Lowe, 1975, 1976) and move as a so-called fluidized sediment flow. Lawson (1979a, 1982) demonstrated in his subaerial studies that grain flows were nonexistent, and that in the field it was almost impossible to make sharp distinctions between some of these flow processes on the basis of quantitative criteria such as water content and shear strength (Figure 7.2). Rapid fluctuations in water content tended to control the mechanism of sediment transport at any given time of observation.

Under these circumstances, it is worth observing that any of these processes may be a precursor for turbidity currents in marine environments and no intermediate stages are required. Thus, according to Heezen and Ewing (1952), the Grand Banks turbidity slump, triggered by an earthquake, passed laterally into a turbidity current without passing through debris-flow or grain-flow stages. However, both debris flows and turbidites evolving from slumps were correlated to periodic earthquake events in the Calabrian Ridge of the eastern Mediterranean and attributed to major earthquakes characterized by periodicities of 1,500 years (Kastens, 1984).

TURBIDITY CURRENTS

DEFINITION AND DEPOSITIONAL MECHANICS

A turbidity current is a turbulent mixture of sediment and water that is more dense than the ambient fluid through which it moves. It moves downslope by gravitational acceleration. The sediment support mechanism during transport is fluid turbulence. In lakes and oceans, turbidity currents move as surges, being initiated by earthquakes, slope failure, or gradation from one of the four previously mentioned sediment gravity processes. The life span of a turbidity current is a function of the sediment volume that is entrained.

A turbidity current can be subdivided into a surging head, a thinner neck behind the head, and a body, which is thicker than the neck. Sediment is moved into the head, although deposition takes place behind the head. When all the sediment is deposited, the current is dissipated. This process of deposition was observed experimentally by Middleton (1966a,b, 1967) and confirmed by others.

In the marine environment, turbidity currents appear to occur and originate within submarine canyons. They also erode and form such canyons. Turbidity currents flow down canyon and then are channelized into submarine fans. Fan growth and aggradation occurs by channel extension in much the same way that a delta distributary extends and builds a delta. The turbidity currents appear to spread as thin sheets on the distal toes of submarine fans and adjoining basins. Transport distances of 4,000 km have been documented (Chough and Hesse, 1976).

Komar (1969, 1972) demonstrated that the flows are channelized and that the interchannel regions of a fan may receive sediment by overbank spill. In regions of higher slopes, flow is supercritical and overbank sedimentation occurs by rapid expansion of the head zone, and head spill. This process would tend to occur in upper fan zones. Downslope, the flow becomes subcritical as slope angles decrease, and overbank deposition occurs by body spill. This process would tend to occur in both the lower regions of inner fan environments and mid-fan settings.

In addition to being fed from the body and neck region, the head of a turbidity current may also act as a zone of erosion, with deposition immediately behind the head. Excess turbulence in the head may produce scour marks and sole marks, which are filled rapidly.

The support mechanism for turbidity currents is fluid turbulence. The gravity-driven motion of the fluid supplies excess turbulence, which suspends the sediments. When these factors of gravity, suspension, and turbulence are in equilibrium, the current system is considered to be in a state of *autosuspension.*

Deposition of sediment by turbidity currents is very rapid, particularly if a sudden slope reduction occurs, such as at the foot of a submarine canyon. Deposition may be rapid where there is a reduction in the degree of turbulence, or dissipation of energy by overbank, head, or body spill. The depositional mechanism is one of particle-by-particle settling.

The textures and structures occurring in turbidites (sediments deposited by turbidity currents) have been documented by many workers, and are organized into a fining-upward sequence (Bouma, 1962), which is shown in Figures 7.6 and 7.7. This sequence, referred to as the "Bouma sequence," is characterized by a sharp basal scour with sole marks on the base, including flute marks produced by turbulent scour, groove casts produced by dragged objects, and various types of prod, bounce, and skip marks. This basal scour surface is overlain by a graded bed (*A* interval), which is in turn overlain by a parallel-laminated interval (*B* interval). Next above is the micro-cross-laminated fine-grained sandstone interval (*C* interval), and it is overlain by an interval of interbedded parallel-laminated, siltstone and claystone (*D* interval). The entire sequence is capped by a pelagic clay. Load deformation structures, pull-apart structures, and slump folds may also occur. Convolute bedding is common and

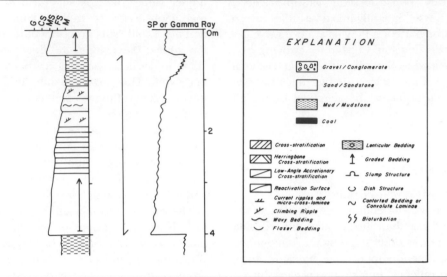

FIGURE 7.6. *Bouma sequence and associated log motif. Graded interval is referred to as A interval, parallel-laminated interval as B interval, micro-cross-laminated and climbing ripple is C interval, interbedded silt and clay zone is D interval, and clay at top is E interval. Doubly terminated arrow shows complete sequence. (Abbreviations: G—Gravel; CS—Coarse sand; MS—Medium sand; FS—Fine sand; M—Mud.)*

may comprise part of the *C* interval of the Bouma sequence.

The mode of formation of the Bouma sequence is considered generally to be one of progressive deposition by a decelerating turbidity current that was reducing slowly its velocity and associated competency and capacity (Bouma, 1962; among others). The process of deceleration involves a change from upper flow (or supercritical) flow conditions (*A* and *B* intervals) to lower regime conditions (*C* interval). More recently, however, it has been suggested by Taira and Scholle (1979) from a study of the magnetic and grain orientation of several Bouma sequences, that the *A* interval is deposited by debris-flow processes that passed into turbidity currents, which deposited the *B, C,* and *D* intervals. The grain fabric of the *A* interval is diffuse and random, whereas in the remaining intervals it is consistent. This diffuse, random orientation of the sand grains in the graded interval is consistent with Lawson's (1979b) observation of random orientation of gravel clasts in debris-flow deposits. Perhaps, the Bouma sequence records a history of processes more complex than previously suggested.

HOLOCENE TURBIDITES

The best documentation of subaqueous gravity deposits, including turbidites, has come from a variety of studies of

sediments dealing with continental slope processes, particularly those involving cable breaks. Extensive Holocene slides and slumps were reported by several workers from continental margin settings, including McGregor (1977), Keller, Lambert, and Bennett (1979), Booth (1979), Hampton and Bouma (1977), among others mentioned earlier. These slides were documented by means of seismic profiling (Moore, 1977). Submarine debris flows are also known; one of the best documented was reported by Embley (1976) off West Africa, where one flow transported sediment a distance of 800 km.

FIGURE 7.7. *Complete Bouma sequence in volcaniclastic sandstone showing basal graded bed (Ta), overlain by parallel-laminated sand (Tb), micro-cross-laminated interval (Tc), parallel-laminated siltstone (Td), and pelagic mudstone (Te). Site 313, Core 38, Section 3 (72–117 cm), northern edge, Mid-Pacific Mountains, Pacific Ocean (20°10.52′N; 170°57.15′W).*

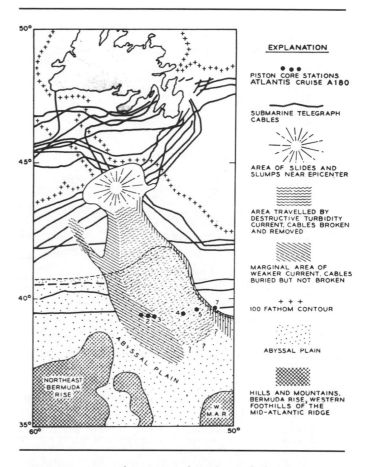

FIGURE 7.8. *Map of 1929 Grand Banks Turbidity Current (from Heezen, 1963; republished with permission of John Wiley & Sons, Inc.).*

The Grand Banks Earthquake of 1929 (Heezen and Ewing, 1952) off southern Newfoundland triggered both a large-scale regional slump and a turbidity current (Figures 7.8 and 7.9) in the eastern fan valley of the Laurentian Fan (Gardner et al., 1984). This area was traversed also by transatlantic telephone cables, each of which was broken during this earthquake by a combination of the aftershocks of the earthquake and by slides and turbidity currents triggered by the earthquake. The times of the cable breaks are known and, as shown in Figure 7.9, using both the time of the break and the distance of transport, it is possible to estimate the velocities of the turbidity cur-

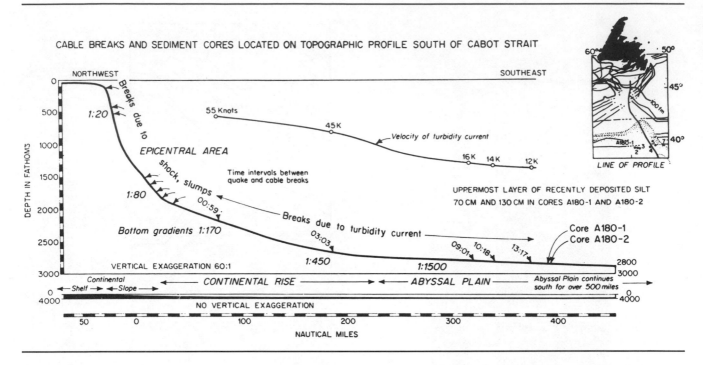

CABLE BREAKS AND SEDIMENT CORES LOCATED ON TOPOGRAPHIC PROFILE SOUTH OF CABOT STRAIT

FIGURE 7.9. *Profile over Grand Banks, Newfoundland, turbidity current showing position of cable breaks and velocity gradient of turbidity current (from Heezen, 1963; republished with permission of John Wiley & Sons, Inc.).*

rents that broke the telephone cables. Velocities range from 55 nautical miles per hour at the base of the continental slope to 12 nautical miles per hour nearly 300 nautical miles seaward in the adjoining Sohn Abyssal Plain. Comparable velocities were determined from telephone cable-break data associated with the Orleansville, Algeria, Earthquake of 1954 (Heezen, 1963). More recently, anomalously thick turbidites (up to 200 m thick) in basinal plain accumulations that lack channeling were attributed to an identical seismic triggered origin by Mutti et al. (1984), who proposed the term "seismoturbidites" for these sediments. They also tend to be much coarser grained than fan turbidites. They differ from other subaqueous gravity-deposited sediments, such as those triggered by tsunamis (Kastens and Cita, 1981); these tend to be more mixed texturally and are similar to muddy debris flows.

SUBMARINE FANS

The major locus of turbidite deposition is the submarine fan environment, as well as abyssal plains (Heezen, 1963;

Sheridan, Golovchenko, and Ewing, 1974; Whitaker, 1974; Normark, 1969, 1978; Normark, Piper, and Hess, 1979; Normark, Piper, and Stow, 1983; Normark, Barnes, and Coumes, 1984; Normark, Mutti, and Bouma, 1984). Transport of sediment is dispersed down submarine canyons and across submarine fans into abyssal basins beyond the fan. Within the canyons, debris flow and fluidized sediment flow is dominant, with accessory turbidity current activity. On the submarine fans, the flow of turbidity currents is channelized.

Within the past year, the nature of sediment processes, morphology, and sediment characteristics of submarine fan systems have been reappraised completely (Normark, Mutti, and Bouma, 1984; Normark and Barnes, 1984). What has emerged from a comparative study of different submarine fans (Barnes and Normark, 1984) is that each is characterized by distinctive features and generalization

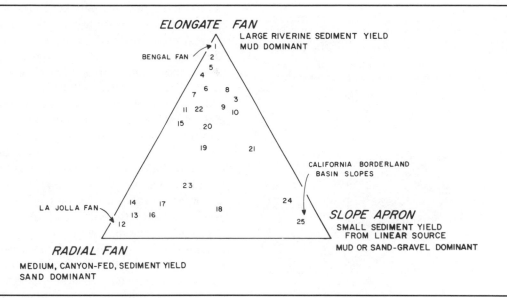

FIGURE 7.10. *Ternary diagram showing variability of present-day submarine fans with respect to three end members consisting of elongate fan, radial fan and a line-source, slope-apron system. (1) Bengal Fan, (2) Indus Fan, (3) Mississippi Fan, (4) Zaire Fan, (5) Amazon Fan, (6) Reserve Fan, (7) Rhone Fan, (8) Laurentian Fan, (9) Monterey Fan, (10) Delgada Fan, (11) Astoria Fan, (12) LaJolla Fan, (13) Redondo Fan, (14) Navy Fan, (15) Nitinat Fan, (16) Coronado Fan, (17) San Lucas Fan, (18) Ebro Fan, (19) Hudson Fan, (20) Orangedo Fan, (21) Crati Lucas Fan, (22) Nile Fan, (23) Menorca Fan, (24) Normal slopes and rises, (25) California continental borderland basin slopes (redrawn from Stow, Howell, and Nelson, 1984).*

of a model and sequence of fan deposition may well be premature. Fans can be either sand dominant, mud dominant, elongate, or radial. Fan systems can develop from point sources at the mouths of submarine canyons or from line sources along the continental margin (Stow, Howell, and Nelson, 1984; Schlager and Chermak, 1979). Nevertheless, certain common characteristics permit the recognition of three fan end members (Figure 7.10): elongate fans, radial fans, and slope aprons. Elongate fans generally occur in areas of large riverine sediment yield and tend to be mud dominant, whereas radial fans tend to be characterized by an intermediate volume of sediment yield provided by submarine canyons and are sand dom-

inant. Slope apron accumulations tend to develop from line sources with a small sediment yield, and are characterized either by mud, or a mixture of sand and gravel. Most Holocene submarine fans are intermediate between these end members, although the Bengal, Indus, and Amazon fans are typical elongate fans, whereas LaJolla, Redondo, Navy, and Coronado fans are typical radial fans. The fanlike basin fills of the California Borderland are characteristically slope aprons (Figure 7.10). These fan systems show as much, if not more, variability and complexity as Holocene deltas (Galloway, 1975; Coleman, 1976, 1980; Normark, Mutti, and Bouma, 1984; Normark and Barnes, 1984).

Because of the large degree of variation between separate submarine fan systems, it seems more appropriate to review selected examples of submarine fans separately rather than to characterize the submarine fan system in terms of one or two well-described fans. The submarine fans to be described below are each different, forming under both active and passive continental margin conditions, different climatic regimes, and different basinal settings. The submarine fans selected to illustrate these differences are the Navy, Mississippi, Amazon, Laurentian, and Bengal deep sea fans.

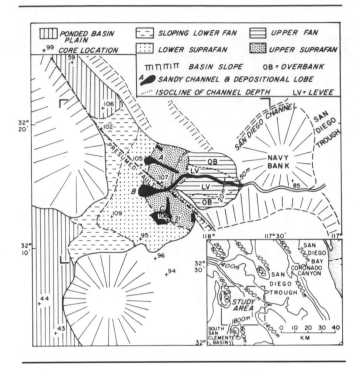

FIGURE 7.11. *Major morphological features of Navy Fan, California (redrawn from Piper and Normark, 1983).*

NAVY FAN

Navy Fan is probably the best-studied submarine fan and has served as a standard for understanding the growth of submarine fans since 1969 (Normark, 1969, 1978; Normark and Piper, 1972). In many respects, it has become a standard for submarine fans in much the same way as the delta of the Mississippi River became a reference standard for deltas.

Navy Fan occurs in the California Borderland region within the San Clemente Basin (Figure 7.11). The basin is floored by an irregular topography because it is tectonically active. The fan has prograded into both an elongate trough and into the basin itself. It is subdivided into several subcomponents. The upper fan is characterized by high levees and steep channels occurring at the mouth of Navy Channel. The mid-fan contains channels that are somewhat shallower because of a decrease in slope gradient and are broken into a series of distributary channels

at the mouth of which smaller lobes or suprafans occur. Beyond this zone of suprafans, which may be amalgamated, is an outer or lower fan zone that lacks channels. The largest concentration of sand occurs within the suprafan lobes and channels (Normark, 1969, 1978; Normark, Piper, and Hess, 1979; Normark and Piper, 1972, 1984; Piper and Normark, 1983; Bowen, Normark, and Piper, 1984). Fan growth appears to have developed by a continuous yield of sediment from Navy Channel causing a continuous progradation of the fan during the Pleistocene and Holocene. The original model of fan growth that is widely cited was one of a simple progradation of inner fan over mid-fan, mid-fan over outer fan, and outer fan over basin plain. Growth is controlled by channel pattern (Normark, Piper, and Hess, 1979).

Sand is concentrated into the channel floor and on the suprafan lobes confined to the mid-fan region. Turbidity currents are channelized (Komar, 1969, 1972) and extend through the main and distributary channels. Because channel relief decreases and terminates in the mid-fan region, sand is dispersed as lobes beyond the distributary channel mouths. Sandy lobe development on the mid-fan tends to accumulate with continuous sediment yield, but because of local tectonics and continuous accumulation, the lobes themselves may reach a stage where channel flow is diverted around the edges of the sand lobes and the lobes are abandoned. Such abandonment causes the channel systems to bend, extend themselves, and deposit new lobes (Figure 7.12).

Sand accumulation is aided by a fundamental mode of turbidity current hydraulics. Komar (1972) established long ago that turbidity currents can grow in height greater than the channels to which they are confined and deposit sediments laterally away from the channel by mechanisms either of head spill in the upper fan, or body spill in the middle fan. More recently, Piper and Normark (1983) and Bowen, Normark, and Piper, (1984) recognized that this process causes a separation of channelized turbidity currents into two parts—that part confined to the channel and that part dispersed into the interchannel zone of a fan. They termed this process "flow stripping" (Figure 7.13), and it tends to be favored at major bends in channels on the fan. Moreover, 80% of the Holocene turbidites on Navy Fan are deposited from the uppermost part of channelized turbidity currents in the interchannel

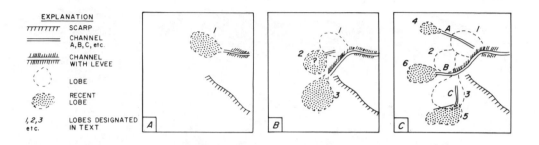

FIGURE 7.12. *Developmental sequence of present channel and depositional lobe pattern, Navy Fan (redrawn from Normark, Piper, and Hess, 1979).*

regions (Piper and Normark, 1983). Flow stripping can involve both head spill (Chough and Hesse, 1980) and body spill processes (Piper and Normark, 1983). Only minor amounts of suspended sand, transported as suspended load, are dispersed away from the channels by flow stripping processes. The lateral extent of different aspects of flow-stripped turbidity currents is shown in Figure 7.13.

ASSOCIATED FANS, CALIFORNIA BORDERLAND

LaJolla Submarine Fan. The LaJolla Submarine Fan occurs also off southern California. Sediment yield onto the fan is through a combination of littoral drift along shore to the head of LaJolla Canyon, sand flow down canyon, and cliff retreat along shore (Shepard, Dill, and Von Rad, 1969). A fourth source of sediment onto the fan is by means of a complex route involving a major sea valley (Loma Sea Valley) that extended landward by cutting into the continental rise and heading into existing minor canyons; this mode of headward canyon cutting is not unlike a process of stream piracy (Graham and Bachman, 1983).

Sediment transport and resedimentation processes in this canyon-fan setting include turbidity currents, storm surges, and normal bottom currents characterized by a tidal component (Shepard, Dill, and Von Rad, 1969; Shepard, 1976; Shepard and Marshall, 1973; Piper, 1970). As at Navy Fan, sand is confined primarily to channels on the fan; no distinct lobes were mentioned by prior workers, although they acknowledge that the largest volume of sand appears to be confined to the mid-fan

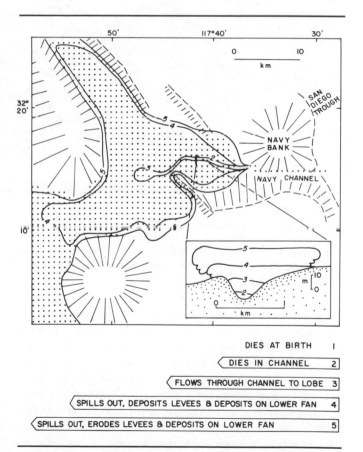

FIGURE 7.13. *Distribution on Navy Fan of thickness and extent of turbidite deposition within channels and those accumulating by turbidity-current flow stripping in overbank settings (redrawn from Piper and Normark, 1983).*

FIGURE 7.14. *Routes of sediment dispersal from river mouth, through longshore currents into submarine canyon by gravity flows and turbidity currents to submarine fan on basin floor. Solid arrows* show system of sand dispersal; *dotted arrows* show mud dispersal *(from Moore, 1972; republished with permission of the Geological Society of America).*

zone, where it tends to be buried by the products of slumping and slides. The morphology of the fan is controlled structurally (Shepard, Dill, and Von Rad, 1969; Graham and Bachman, 1983).

Monterey Fan and Delgada Fan. The Monterey Fan occurs in relatively deep water off the north-central California coast and is deposited on oceanic crust of late Oligocene age (Normark et al., 1984; Wilde, Normark, and Chase, 1978). It is zoned into an upper fan, mid-fan, and outer fan region that is similar to Navy Fan. Turbidity currents are responsible for developing sand bodies on the fan; these sand bodies are in the form of either large valley fills or suprafan lobes. Only two suprafan lobes were reported, however (Wilde, Normark, and Chase, 1978). Sand bodies are buried, presumably by slump processes.

Monterey Fan is receiving sediment from two canyon systems, the Monterey and Ascension canyons. These canyons are characterized by a complex history involving canyon piracy. Other abandoned valleys are also known in the continental slope region adjoining the fan (Normark et al., 1984).

Delgada Fan occurs immediately north of Monterey Fan and coalesces with it (Normark and Gutmacher, 1984). Fan development began in late Miocene time (Hein, 1973; Normark and Gutmacher, 1984). Because of its location on the Pacific Plate, Delgada Fan has moved northwest, and as a consequence its mineralogy has changed through time as different terrigenous sources provided sediment onto the fan (Hein, 1973). Most sand accumulated in the major channels occurring on this fan and some accumulated seaward of a lobate fan-valley complex. No suprafan-type lobes were detected in preliminary surveys.

Redondo Submarine Fan. The Redondo Submarine Fan is located in the northern end of the San Pedro Basin in southern California. It differs from the descriptions of the other California fans mentioned so far because it has been segmented by faulting, which causes a change in channel gradients and associated sediments (Haner, 1971). Some of the channels are entrenched onto parts of the fan surface, and laterally, sediment properties tend to change abruptly and are placed in contact with sediment distributions that are in equilibrium with a pretectonic, or prefault movement condition. Turbidity currents are confined primarily to channels on the upper fan segment, whereas in the lower and middle fan regions, turbidity currents are not confined to channels at all and spread in a sheetlike mode. Major shifts in channels tend to occur in response to faulting. As a consequence, predicting sand occurrences in a potential counterpart becomes difficult because promising sand bodies are cut off by active faults, placing many outer and middle fan styles of sedimentation against inner fan sand bodies, for instance.

DISPERSAL OF SEDIMENT ONTO CALIFORNIA SUBMARINE FANS

Sediment dispersal to the submarine fans of coastal California involves subaqueous gravity processes eroding canyon walls, littoral drift along shore by longshore currents, and riverine sediment yield. In the California Borderland, Moore (1972) demonstrated that the sediments on submarine fans are derived from terrigenous sources involving fluvial transport to a coastline where longshore current systems transport coarser sizes along shore. Because the submarine canyons of this region head close to shore (Figure 7.14), they act as a trap for this sediment and divert it downslope by gravity processes that change to turbidity currents with a progressive increase in fluid content and turbulence. These currents then develop the submarine fan morphology and sediment distribution patterns reviewed above (Figure 7.14).

AMAZON DEEP-SEA FAN

The Amazon Deep-Sea Fan extends from the continental shelf off northeastern Brazil. It is the third-largest Holo-

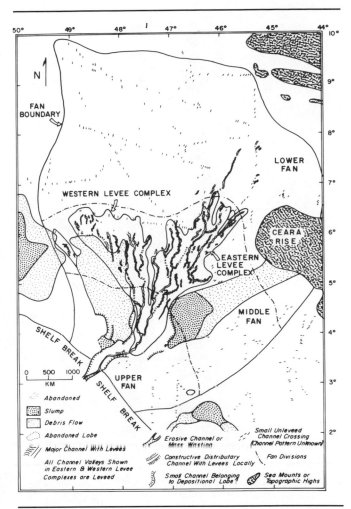

FIGURE 7.15. *Amazon Fan morphology showing location of major channels and debris-flow areas of major levee complexes (redrawn from Damuth and Flood, 1984).*

cene submarine fan in the world (Damuth and Flood, 1984). It extends from the shelf break of the Amazon continental shelf (elevation of 400 m below sea level) to a depth of 4,900 m over a distance of 700 km (Figure 7.15). This fan is divided into three zones, an upper fan, a middle fan, and a lower fan. The upper fan is characterized by a rugged topography with steep scarps and a gradient averaging 14 m/1,000 m. The upper fan is cut by the Am-

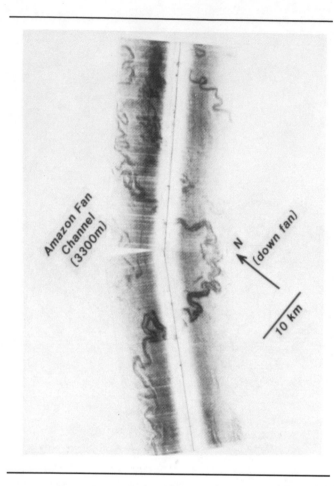

FIGURE 7.16. *Computer-enhanced GLORIA sonograph of meandering channel in Amazon Fan (sonograph courtesy of R.D. Flood; republished with permission of R.D. Flood).*

azon Submarine Canyon, which contains a major channel system that splits into several distributary channels (Figure 7.15). The rugged topography in the upper fan zone is attributed to extensive mass transport of material by both slumps and debris flow (Damuth and Embley, 1981).

The middle fan is characterized by an average gradient of 5 m/1,000 m, and a relatively smoother topography. Its diagnostic feature is the presence of extensive channels (Damuth and Kumar, 1975; Damuth and Flood, 1984), which break up into a series of distributaries. These channels, moreover, are demarcated from the interchannel fan region by a complex set of extensive levees (Figure 7.16). Individually, these channels show a pronounced meandering pattern (Figure 7.16; Damuth, Kolla, Flood, Kowsmann, Monteiro, Gorini, Palma, and Belderson, 1983), some with a very pronounced and exaggerated sinuosity. Channel bifurcation was observed on the midfan zone and appears to develop from levee breeching. The sinuosity of these channels suggests also that turbidity-current flow in these channels is a continuous, rather than an intermittent phenomenon (Damuth, Kolla, Flood, Kowsmann, Monteiro, Gorini, Palma, and Belderson, 1983).

The outer fan is characterized by a low gradient, averaging 2.3 m/1,000 m, and its surface is incised by small distributary channels lacking levees. Acoustically, this zone is layered and distinct, whereas the upper and middle fan zone is characterized by acoustically transparent zones (Damuth and Kumar, 1975; Damuth and Embley, 1981; Damuth and Flood, 1984).

The Amazon Fan is characterized by a history of growth that differs in part from the original submarine fan growth model proposed by Normark (1969). In the Amazon Fan, growth appears to involve both lateral progradation and vertical accretion (Damuth, Kousmann, Flood, Belderson, and Gorini, 1983). Vertical accretion is dominated by the distributary channel systems that cause vertical growth by formation of a succession of channel-levee complexes. Channels appear to be active only at a given time and are abandoned by channel avulsion and development of a younger and new channel-levee complex. The channel-levee complexes coalesce and overlap each other as new channel-levee complexes adjoin abandoned channel complexes, causing a succession of such channel and levee systems to form. Evidence for this vertical growth comes from analysis of both seismic and side-scan sonar (GLORIA) records (Damuth, Kousmann, Flood, Belderson, and Gorini, 1983).

Most of the sand in the Amazon Fan is confined to the distributary channel and levee complexes of the mid-fan zone. The overlapping distributary channel-levee complexes that account for fan growth would give rise to a series of discontinuous sand bodies.

MISSISSIPPI FAN

The Mississippi Fan is located in the east-central part of the Gulf of Mexico and consists primarily of a large volume of Pleistocene sediment overlain by a thin veneer of Holocene sediments (Bouma, Stelting, and Coleman, 1984; Moore et al., 1978; Leg 96 Scientific Staff, 1984). It is bounded by three major escarpments on the sea bed (Florida, Sigsbee, and Campeche) and the Sigsbee Abyssal Plain. This fan contains 290,000 km³ of sediment and extends over an area exceeding 300,000 km² (Bouma, Stelting, and Coleman, 1984, p. 147). The fan has been divided into three zones similar to other fans, namely the inner, middle, and outer fan (Moore et al., 1978; see also Figure 7.17).

The Mississippi Fan heads into the Mississippi Canyon (Figure 7.17), whose head occurs in turn about 50 km southwest of Southwest Pass on the Mississippi Delta. This canyon formed about 25,000 to 27,000 BP (Bouma, Stelting, and Coleman, 1984, p. 149) by a combination of slumping and sliding, probably in response to a lowering of sea level. Continued slumping along this canyon appears to account for the bulk of sediment dispersed onto the Mississippi Fan, with lesser volumes of sediment being provided from terrigenous sediment yield dispersed through the canyon.

A combination of seismic surveys and side-scan sonar mapping, using the GLORIA system, was conducted as part of surveys in conjunction with Leg 96 of the Deep Sea Drilling Project (Garrison, Kenyon, and Bouma, 1982). These surveys demonstrated that the Mississippi Fan consists of a series of amalgamated lobes that were active at different times. Their positions were controlled by relief of older fan lobes and accumulation in topographic lows. Eight such lobate units were defined by seismic mapping. An additional factor controlling the position of these lobes

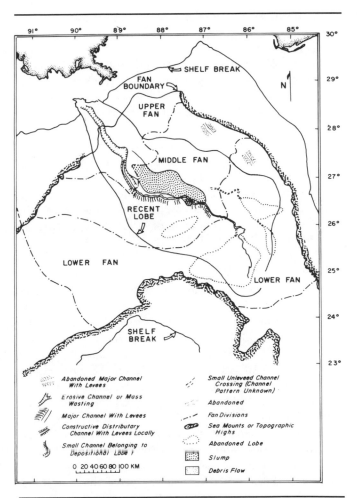

FIGURE 7.17. *Mississippi Fan morphology showing location of major slump system and major meandering channel system (redrawn from Bouma, Stelting, and Coleman, 1984).*

is the position of the major source point of sediment yield into the basin, either by slumping, terrigenous-derived transport, or both.

Each lobe sequence associated with individual fan lobes appears to have accumulated in response to a time of rising sea level (Bouma, Stelting, and Coleman, 1984) and was initiated with slumping triggered by a low stand of sea level.

Processes providing sediment onto the fan are variable. The dominant processes appear to be shelf-edge slumping combined with extensive slumping and erosion of the walls of the Mississippi Canyon. Recognizable slump features in seismic sections confirm the existence of these features in the inner fan zone of Moore et al. (1978) and can be traced back to their original scars (Bouma, Stelting, and Coleman, 1984). Density currents appear to be infrequent. Bouma, Stelting, and Coleman (1984) postulated that the major slumps and slides were progressively fluidized into debris flows and, ultimately, into turbidity currents. However, these processes would occur down fan from Moore et al.'s (1978) inner fan zone.

The middle portion of the most recent fan lobe is characterized by an extensive meandering channel system (Garrison, Kenyon, and Bouma, 1982; Bouma, Stelting, and Coleman, 1984; Leg 96 Scientific Staff). Abandoned channel systems were also observed. Lateral migration of the channels is suggested not only from a preserved ridge-and-swale topography, preservation of overbank deposits and levees, and crevasse splays (Bouma, Stelting, and Coleman, 1984), but also from the preservation of fining-upward sequences in DSDP sediment cores (Leg 96 Scientific Staff, 1984). This interpretation assumes deposition and aggradation in fluvial terms. However, as demonstrated by Damuth and Embley (1984), channels on the Amazon Fan, which are similar to those in the Mississippi Fan, require continuous flow of turbidity currents, an interpretation consistent with known channelization of turbidity flows (Komar, 1969). Bouma, Stelting, and Coleman (1984) acknowledged this possibility as a second alternative, indicating that flow stripping would provide thin turbidites on the interchannel zones and cause levee accumulation. Later drilling (Leg 96 Scientific Staff, 1984), confirmed this second alternative. These channels reach a width of 3 to 4 km and a relief of 50 m on the middle portion of the fan, but decrease to a width of

500 m and a relief of 10 m in the lower part of the fan (Leg 96 Scientific Staff, 1984).

Sediment accumulation rates were extremely large, averaging 11.75 m per 1,000 years in the middle portion of the fan, and 6.46 m per 1,000 years on the outer part of the fan. Sand is known to accumulate not only in channel complexes, but also in smaller lobate bodies seaward from the fan-channel mouths.

LAURENTIAN FAN

The Laurentian Fan is located seaward from the Laurentian Channel, an incised linear drowned glacial trough cut into the Atlantic continental shelf of Canada. The Laurentian Channel provided a transport path for both ice and glacially transported sediment during lower stands of sea level during the Quaternary. This sediment is the principal source of material from which the Laurentian Fan appears to have been built (Piper, Stow, and Normark, 1984; Stow, 1981; Normark, Piper, and Stow, 1983; Gardner et al., 1984).

The Laurentian Fan extends from the seaward end of the Laurentian Channel to the Sohn Abyssal Plain (Figure 7.18). It is an elongate fan and its diagnostic characteristic is that the major proportion of its sediment consists of mud. Sand appears to be limited exclusively to channels and sandy depositional lobes, particularly in the middle of the fan.

Physiographic zonation of the Laurentian Fan has been arbitrarily divided by Stow (1981) into the well-known scheme of Normark (1969) of inner, middle, and upper fan. However, the subdivision was abandoned (Piper, Stow, and Normark, 1984) and more descriptive morphological units characteristic of this fan were proposed (Figure 7.18). The first of these is the continental slope, which is characterized by irregular relief, presumably as a result of erosion of Quaternary sediment by slope processes. The slope extends below the shelf break to a depth of about 2,000 m. The slope-valley transition occurs seaward of the continental slope and consists of complex erosional valleys containing slide blocks and intervalley accumulations of sediment. This slope-valley transition zone also acts as a line source for the bulk of sediment accumulating on the Laurentian Fan. Some of this sediment was dispersed during the 1929 Grand Banks Earthquake (Gardner et al., 1984). It grades downslope into a channel-levee complex cut by two fan valleys with asymmetric levees. These levees are Late Quaternary in age. This zone grades laterally into an area where valleys terminate, bifurcate and are characterized by smaller relief. The lowest end of the fan, which is perhaps the largest zone, is an area of sandy depositional lobes and is distinguished from the other areas by lacking major channels. A few shallow channels occur, however.

Slumping of glaciomarine sediment and till from the upper continental slope comprises the major source for sediment on the Laurentian Fan. Much of this is triggered seismically (Gardner et al., 1984). Gravels and sands occur within the fan valleys, and sand is characteristic also of the sandy depositional lobe. Large debris flows occur along the eastern part of the upper part of the Laurentian Fan. Turbidity currents are responsible for transporting and redistributing sand within the sand depositional lobe zone as well as within channels. Resedimentation by contour-hugging bottom currents redistributes some of the sediment and also is responsible in part for burying sandy turbidites and the seaward extension of channel deposits.

BENGAL DEEP-SEA FAN

The Bengal Fan occurs in the Bay of Bengal between the Indian subcontinent and the Ninety-East Ridge. It is the largest submarine fan known in the world with areal dimensions of 2,800 to 3,000 km in length and a maximum width of 1,430 km (Emmel and Curray, 1984). Gradients on the fan range from 6 m/km in the innermost part to less than 1 m/km in its outermost part (Figure 7.19).

This fan is subdivided into an upper fan, mid-fan, and lower fan. The upper fan is characterized by deeply incised valleys with an average gradient of 2.39 m/km. These valleys contain one or more terraces, well-developed levees, widths on the order of 10 km, and a thalweg higher than adjoining fan surfaces. Such valleys are connected to their source and incise depositional lobes. In the mid-fan, valleys are connected to upper fan valleys, are v-shaped in section, contain an occasional narrow terrace, and are still active. They extend into parts of the lower fan region. A third type of valley, common to both the mid-fan and lower fan, is characterized by wide flat bottoms, smooth rounded levees if present, and terraces, and represents abandoned channel systems. Some parts of

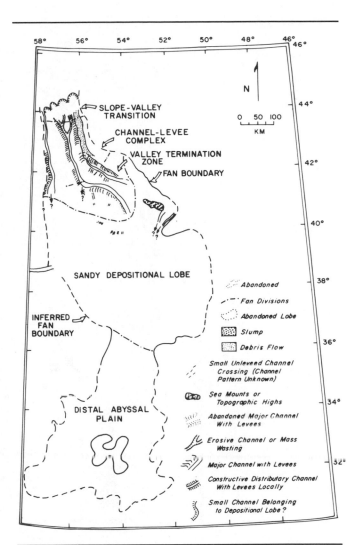

FIGURE 7.18. *Laurentian Fan morphology showing major channel and levee complexes, fan valleys, and depositional lobes (redrawn from Piper, Stow, and Normark, 1984).*

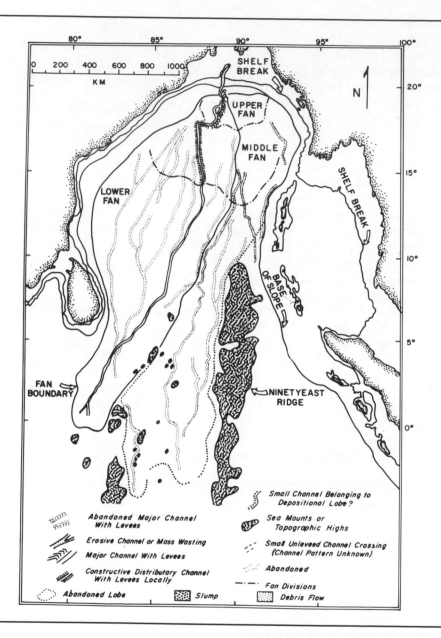

FIGURE 7.19. *Bengal Fan morphology, showing position of major channel systems (redrawn from Emmel and Curray, 1984).*

these third types of channels may be rejuvenated, in which case paired terraces occur (Emmel and Curray, 1984).

The valleys appear to be the main conduit of transport of gravel and coarse and medium sand (Curray and Moore, 1971; Emmel and Curray, 1984). Many of these valleys are buried and observed in seismic records (Figure 7.20) where they are demarcated by acoustically transparent zones that are diagnostic of such channels. The levees of Pleistocene valley-fills also appear on many seismic records of upper fan valleys. Three cycles of sand were reported during drilling of DSDP Site 218, with the Pleistocene cycle being 60 m thick (Von Der Borch, Sclater et al., 1974). A thin layer of Holocene mud was found to mantle much of the present Bengal Fan.

Sediment influx to the Bengal Fan is correlated to rates of denudation of the Himalayan Mountain Chain since middle Miocene time (Curray and Moore, 1971; Emmel and Curray, 1984; Seeber and Gornitz, 1983). The volume of seismically defined units in the Bengal Fan seems to correlate well with known rates of sediment yield in the Ganges-Brahmaputra Delta (Curray and Moore, 1971), indicating that rate of uplift is the dominant control in sediment accumulation in this particular fan.

SUMMARY OF HOLOCENE SUBMARINE FANS

The above brief review of submarine fans is a partial compilation of current knowledge of submarine fans. Other examples of relatively well-described fans include Astoria Fan off the coast of Oregon (Nelson, 1984; Nelson and Kulm, 1973; Nelson and Nilsen, 1974; Carlson and Nelson, 1968; Nelson et al., 1970), the Crati Fan in the Ionian Sea (Ricchi Lucchi et al., 1984), the Magdelena Fan off Colombia (Kolla and Buffler, 1984; Kolla, Buffler, and Ladd, 1984), the Ebro Fan of southeastern Spain (Nelson et al., 1984), the Indus Fan (Kolla and Coumes, 1984), and the Rhone Deep-Sea Fan (Normark, Barnes, and Couma, 1984). These fans and the fans reviewed above are all characterized by a decrease in slope from apex to toe, incision by complex arrays of channels that are acoustically transparent in seismic sections (Figure 7.20), changing proportions of sand and mud, with the Bengal Fan being perhaps sand dominated in contrast to the mud-dominated Laurentian Fan, and complex histories of growth, partly controlled by local depositional events such as in the Amazon, major uplift and denudation, as in the Bengal Fan, or progradational growth, as exemplified in the Navy Fan. In most of these fans, debris flow and slump processes are common and their sediments tend to be preserved within the inner fan zone. Most workers tend to subdivide fans in terms of inner, middle, and outer fan zones, although there is a tendency to move towards a more detailed descriptive subdivision, as shown by Piper, Stow, and Normark (1984) in the Laurentian Fan, for instance. These fans differ considerably in size (Barnes and Normark, 1984). This review reinforces Normark and Barnes' (1984) finding that no one fan can serve as a reference standard for facies modeling, and proposals for such models may be premature. In the past, Mutti and Ricchi Lucchi (1972) suggested that submarine fan sequences should coarsen-upward provided fans are subjected to a progradational growth model such as Normark (1969) proposed earlier for Navy Fan. Confirmation of Mutti and Ricchi Lucchi's (1972) model came from their study of ancient rocks, not modern submarine fans. None of the current workers have suggested a particular fan sequence that shows overall upward-coarsening of texture or increasing bed thickness. Even cursory examination of lithologic logs from DSDP drill sites in the Mississippi Fan (Leg 96 Scientific Staff, 1984, p. 16) fails to indicate such a widespread pattern. No such trend was observed at DSDP Site 218 (Von Der Borch, Sclater et al., 1974) either.

In closing this discussion of modern submarine fans and their turbidites, it must be emphasized that turbidites occur elsewhere in marine environments. The abyssal plains are also sites of turbidites, as indicated from coring and from seismic records (Sheridan, Golovchenko, and Ewing, 1974; Creager, Scholl et al., 1971; Benson, Sheridan et al., 1978; Tucholke, Vogt et al., 1979; among others). There, the turbidites are thin, graded, and interbedded with hemipelagic and pelagic muds. They show a thinly layered laterally persistent seismic stratigraphy (see Sheridan, Golovchenko, and Ewing, 1974).

ANCIENT TURBIDITES AND SUBMARINE FANS

Many ancient turbidite sequences have been documented in the rock record and no attempt is made here to compile all examples. A few are listed below to illustrate the range

172

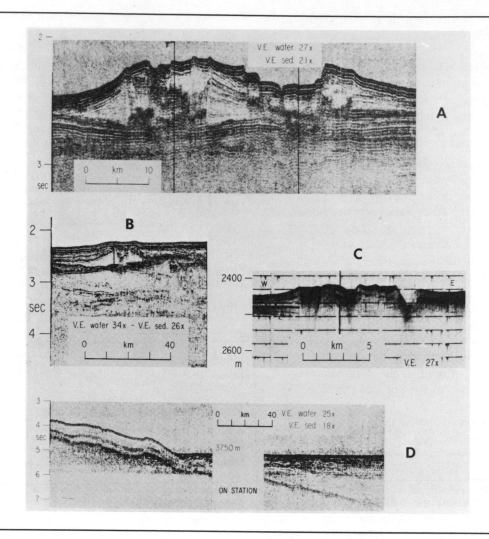

FIGURE 7.20. *Continuous seismic reflection profiles through submarine fan and channel system, Bengal Fan. (A) Reflection profile of large fan valley and natural levee near northern apex of fan. (B) Reflection profile of an abandoned and buried fan valley near northern apex of fan. (C) Reflection record of fan valley in mid-fan zone. (D) Reflection profile, west flank, Ninety East Ridge (from Curray and Moore, 1971; republished with permission of the Geological Society of America).*

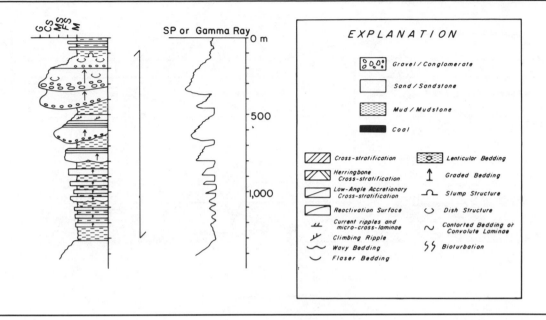

FIGURE 7.21. *Idealized vertical sequence and log motif for ancient submarine fan system. Doubly terminated arrow shows sequence interval. (Abbreviations: G—Gravel; CS—Coarse sand; MS—Medium sand; FS—Fine sand; M—Mud.)*

of features and styles of deposition that comprise this facies of deposition.

The Cenozoic and Cretaceous of the Polish Carpathian Mountains are well-known examples of turbidite formations (Dzulynski, Ksiaskiewicz, and Kuenen, 1959) and the Oligocene–Miocene of Italy (Ten Haaf, 1958; Mutti and Ricchi Lucchi, 1972) comprises other well-known examples. The Ordovician Martinsburg Formation (Mc-Bride, 1962) and the Ordovician Cloridorme Formation (Enos, 1969) of the Appalachians also comprise two well-studied examples. These formations were identified as turbidites from a combination of graded sequences, preservation of Bouma sequences (partial and complete), the occurrence of flute casts, groove casts, and prod and skip marks on the soles of turbidite sandstone beds, and the association of deep-water indicators. Paleocurrent indicators suggested an axial mode of turbidite sediment transport in most of these examples. Other examples include the Pliocene turbidites of the Ventura Basin and the

Butano Sandstone (Eocene) of California (Nelson and Nilsen, 1974; Nilsen, 1984a).

Ancient counterparts of submarine fan systems have been documented also. Mutti and Ricchi Lucchi (1972) in their study of the Oligocene turbidites of the Appenines, recognized several components of submarine fan systems in the units they studied. Using a model of fan growth developed by Normark (1969), they postulated that as fans grow, they prograde basinward away from continental slopes, with slope deposits overlying inner fan sediments, inner fan sediments overlying mid-fan sediments, and mid-fan sediments prograding over outer fan sediments, which in turn prograde over basin sediments. Further, they argued that the overall sequence generated by this process showed a general coarsening-upward succession. In addition, the bed thickness of sandstone units also increases vertically (Kruit et al., 1975). Individual sandstone beds may fine-upward as a Bouma sequence, but the entire fan complex is characterized as an overall coarsening-upward and thickening-upward sequence. Such a generalized submarine fan sequence is shown in Figure 7.21.

In a review of turbidites and submarine fan deposition,

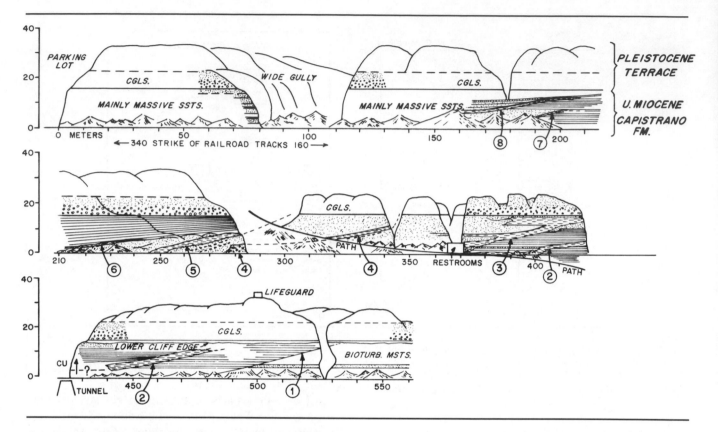

FIGURE 7.22. *Cliff section at San Clemente, California, showing sketch of series of nested submarine channel fills, Capistrano Formation (Miocene) (from R.G. Walker, 1975; republished with permission of the Geological Society of America).*

Walker (1978) emphasized that very few complete examples of submarine fan successions are known or identified, partly because few studies have applied the results of both Normark's (1969) work and the modeling of Mutti and Ricchi Lucchi (1972). One case he cited is the Shale Grit (Carboniferous) of England, which he had identified earlier as a turbidite (R.G. Walker, 1966, 1967, 1973) with channel fills. Other units show evidence of partial preservation of components of submarine fans, and he lists several to which the reader is referred. Perhaps one of the finer examples of a mid-fan suprafan lobe with nested channels is the Miocene Capistrano Formation of California (R.G. Walker, 1975), a cross-section of which is

shown in Figure 7.22. A similar coarsening-upward sequence produced by submarine fan progradation was reported from back-arc basins in the southwest Pacific by Klein (1975c) and from the Eocene of Spain by Kruit et al. (1975).

Another example of a complete coarsening-upward submarine fan succession is represented in the Mississippian Stanley and Jackfork Group of the Ouachita Fold Belt of Arkansas and Oklahoma (Cline, 1960; Cline and Moretti, 1956; Klein, 1966; Briggs, 1974; Morris, 1974a,b; Moiola and Shanmugam, 1984) which the author (Klein, 1980) reexamined at Kiamichi Mountain, Oklahoma. There, the Late Devonian Arkansas Novaculite, representing biogenic pelagic deep-water chert, is overlain by the basal Stanley Group consisting of the interbedded shales and thin sandstones of the Ten Mile Creek Formation. This interval is clearly a basinal facies. Next above is the Moyers Formation consisting of medium-thick fine-grained sandstones and shales, representing the intertonguing of basinal facies and outer fan facies. It is overlain by the Chickasaw Creek Siliceous Shale, another biogenic pelagic deposit, which represents a time during which fan-growth ceased. It is overlain by the Wildhorse Mountain Formation, which contains medium-grained sandstone interbedded with shale. These sandstones also thicken upward and represent not only resumption of fan sedimentation, but also the intertonguing of outer and mid-fan facies. Shallow channels are present in this unit and include partial Bouma sequences (T_{a-b}). It is overlain by the Prairie Mountain Formation, which contains medium- and coarse-grained sandstone, is thick-bedded, and contains interbedded shale, debris-flow conglomerates, and intervals containing slump folds. This unit appears to be an inner fan deposit. Thus, almost the entire Mississippian System at Kiamichi Mountain is represented by the coarsening-upward motif of a probable prograding submarine fan, from the basinal facies upward to outer, mid- and inner fan facies.

More recently, a variety of well-described ancient submarine fan systems were recognized. A partial list includes the Miocene Blanca Fan, California (McLean and Howell, 1984), the Cretaceous of the Chugach Mountains, Alaska (Nilsen, 1984b), the Eocene Ferrello Fan of California (Howell and Vedder, 1984), the upper Cretaceous Chatsworth Formation of California (Link, Squires, and Colburn, 1984), the Gottero Sandstone (Cretaceous–Paleocene) of Italy (Nilsen and Abbate, 1984), and the Eocene Hecho Fan of Spain (Mutti, 1984). Fan deposition was recognized from lateral facies changes from debris-flow conglomerates to intertonguing marine sandstones (McLean and Howell, 1984), external geometry of the example in question, paleocurrents, and recognition of marine channels, including some with preserved evidence of channel migration (Nilsen and Abbate, 1984). Almost none of these reports show evidence of coarsening-upward successions, although subsurface study in one case (Howell and Vedder, 1984) shows a log pattern that compares well to a coarsening-upward and thickening-upward motif. These recent studies may well have considered the admonitions of Normark and Barnes (1984) and downplayed the importance of this sequence, however, which is to be expected.

SEA LEVEL AND SUBMARINE FANS

Recent work by Howell and Vedder (1984) showed that deposition of the Eocene Ferrello Fan of California occurred during a time of a low stand of sea level, using the sea level curves of Vail, Mitchum, and Thompson (1977) and Vail and Hardenbol (1979) as a framework. This finding is consistent with results from modern submarine fans in *passive continental margins.* There, major sedimentation occurs on submarine fans during eustatic lower stands of sea level associated with Pleistocene glaciation (Damuth and Kumar, 1975; Kelts and Arthur, 1981; Damuth and Flood, 1984; Bouma, Stelting, and Coleman, 1984; Piper, Stow, and Normark, 1984; Stow, 1981; among many others). More recently, Shanmugam and Moiola (1982) compiled data concerning the temporal distribution of major Cenozoic submarine canyon and fan accumulations, including several major oil fields, particularly in the North Sea (Figure 7.23). In fact, Vail, Mitchum, and Thompson (1977) appeared to have identified their low stands of sea level on their well-known sea level curve where seismic sections revealed a lateral change from unconformities and other stratigraphic breaks in shelf sequences into fan-shaped packets of sediment that were interpreted to be submarine fans. In *passive continental margins* there appears to be excellent evidence pinpointing times of active sedimentation on submarine fans, presumably because such lowering of sea level increases relief, continental denudation rate, and sediment yield. Also, distance of terrigenous sediment transport from major river mouths would be less to sites of fan accumulation during low stand, than during times when sea level is relatively high.

Caution must be exercised when trying to explore for subsurface submarine fans by assuming maximum fan growth and development during lower stands of sea level. In the Madeira Abyssal Plain of the east central Atlantic Ocean, Weaver and Kuijpers (1983) demonstrated that maximum sediment deposition of turbidites (expressed as preserved frequency of turbidites) occurs not during low stands of sea level, but during times of both transgression and regression on the shelf, rather than a supposed quiet period in between. In *active continental margins* other factors may mask the role of sea level fluctuations. Although Howell and Vedder (1984) reported an ancient active margin submarine fan deposit that accumulated

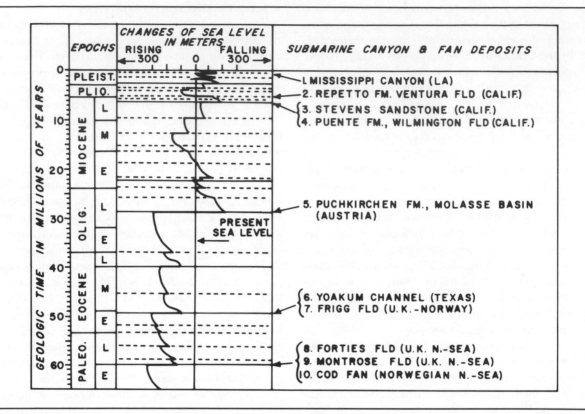

FIGURE 7.23. *Cenozoic sea level curve after Vail, Mitchum, and Thompson (1977) through time and occurrence of major hydrocarbon-bearing submarine canyon and fan reservoirs at time of a low stand of sea level (redrawn from Shanmugam and Moiola, 1982).*

during a low stand in sea level, other examples exist indicating that the role of tectonic uplift may exercise greater influence on fan accumulation. Klein (1984, 1985b) reported that in the Toyama Submarine Fan at DSDP Site 299, the frequency of turbidite deposition increased and periodicity of turbidite deposition decreased during coeval uplift of the Hida Range in Honshu acting as a source. The rate of tectonic uplift of the Hida Range was 1,500 km during the Quaternary, which was ten times the rate of sea level reduction during eustatic changes triggered by glaciation. Other cases where large turbidite frequencies and short turbidite periodicities were observed on submarine fans include DSDP Site 210 in the Coral Sea Basin where maximum accumulation was coeval with rapid rates of tectonic uplift in the Owen-Stanley Range of Papua-New Guinea (Klein, 1984, 1985b). Other cases where submarine fan accumulation occurred at rapid rates independent of sea level fluctua-

tions or during high stands of sea level as per Vail, Mitchum, and Thompson (1977) include the northern Shikoku Basin (DSDP Site 297), and the Hebrides Basin (DSDP Site 296) (Klein, 1985b, in press). These relations appear to be valid if the rate of tectonic uplift exceeds 400 m/million years (Klein, 1985b, in press). Thus, care must be taken in exploring for such targets in active continental margins before applying or modifying existing sea level curves.

OIL FIELD EXAMPLES

Turbidite sand bodies on submarine fan systems comprise one of the four major types of stratigraphic traps

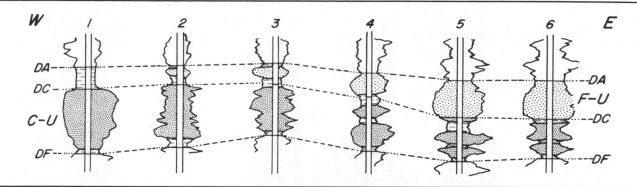

FIGURE 7.24. *Revision of correlation of electric logs by Hsu (1977) of lower Pliocene Repetto Formation, Ventura Basin, California. DA and DC are marker beds. C-U (coarsening-upward sequence); F-U (fining-upward sequence). Shaling out of sandstone units appears to represent lateral pinch-out of suprafan lobes. Wells are (1) Shell-Taylor 505, (2) Shell-Taylor 349, (3) TWA Hartman 44, (4) TWA Lloyd 161, (5) TWA Lloyd 165, and (6) TWA V.L. & W. 84 (redrawn after Walker, 1978).*

from which oil and gas are extracted. A variety of examples are known. Perhaps one of the better-known examples is the lower Pliocene of the Ventura Basin, California (Hsu, 1977; Walker, 1978; Weser, 1975). There, turbidite sand bodies are interbedded with mudstones and comprise part of a submarine fan system (Figure 7.24), primarily suprafan lobes (Walker, 1978). Coarsening-upward deposits (represented by funnel-shaped logs) comprise the main part of the lobe system, whereas the fining-upward log motif represents individual turbidite sandstone beds. Weser (1975) demonstrated that the Los Angeles Basin also shows evidence of a turbidite history of deposition. There, sandstones thicken upward, a feature that shows well in the electric logs recovered from that basin (see Weser, 1975, his Figure 27), suggesting a system of submarine fans.

The log characteristics of submarine fan turbidites are either blunt-base, blunt-top, or transitional base and abrupt top, or funnel-shaped (coarsening-upward), or abrupt base and sloping top (fining-upward). Selley (1976) demonstrated that in the North Sea, the blunt-base, blunt-top pattern and the fining-upward pattern occur in several subsurface oil fields (Figure 7.25). Submarine fan reservoirs are known from the Lower Eocene of Frigg Field of the North Sea Viking Graben (Heritier, Lossel, and Wathne, 1979), where the reservoirs were identified as submarine fans on the basis of isopach mapping of seismic intervals, characterized by acoustically transparent zones (see also Shanmugam and Moiola, 1982).

Cretaceous and Miocene turbidite and submarine fan sandstone reservoirs of petroleum and natural gas have been reported from the Sacramento and San Joaquin valleys of California. The Winters Sandstone (Cretaceous) is part of a submarine fan complex associated with major deltas mapped as the Kione and Starkey formations (Garcia, 1981; Tillman et al., 1981; Williamson and Hill, 1981). Recognition of these reservoirs as submarine channel fills came from analysis of cores showing typical features of turbidites (Tillman et al., 1981; Williamson and Hill, 1981), isopachous mapping of individual sand bodies showing a fan-shaped morphology (Williamson and Hill, 1981; Garcia, 1981), and analysis of logs showing that individual sandstone beds were characterized by fining-upward trends representing partial or complete Bouma sequences. Garcia (1981) demonstrated that the Winters Sandstone assumes a stratigraphic position not unlike the overlapping of deltas over turbidites shown off the Niger Delta by Burke (1972; see also Figure 5.15) and in seismic sections of O.R. Berg (1982; see also Figure 5.32).

The Winters Sandstone (Miocene) of the San Joaquin Valley, California, has been described most recently by Webb (1981) and MacPherson (1978). Again, recognition of fan components came from isopachous mapping (Figure 7.26) of individual sand bodies. These are named

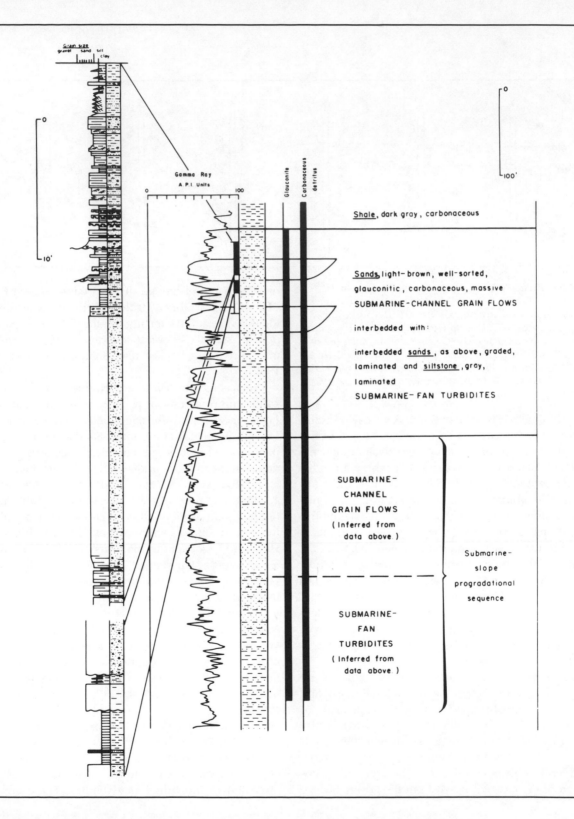

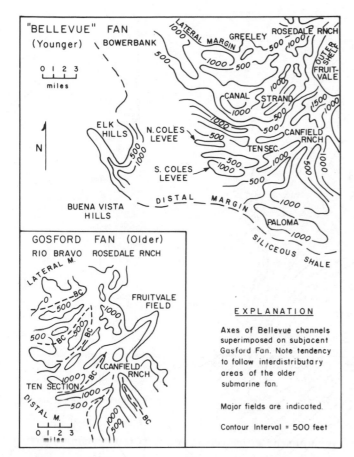

FIGURE 7.26. *Isopach map of Bellevue and Gosford turbidite fans, Stevens Sandstone (Miocene), Bakersfield arch, San Joaquin Basin, California (redrawn from MacPherson, 1978).*

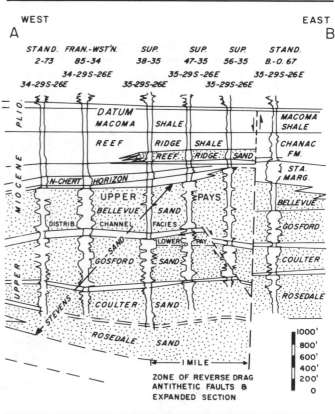

FIGURE 7.27. *Bellevue Field, San Joaquin Valley, California, showing electric log cross-section through several submarine fans in Stevens Sandstone (Miocene). Oil entrapment occurs on downthrown side of growth fault (redrawn from MacPherson, 1978).*

FIGURE 7.25. *Environmental analysis of submarine fan reservoirs in North Sea well (from Selley, 1976; republished with permission of the American Association of Petroleum Geologists).*

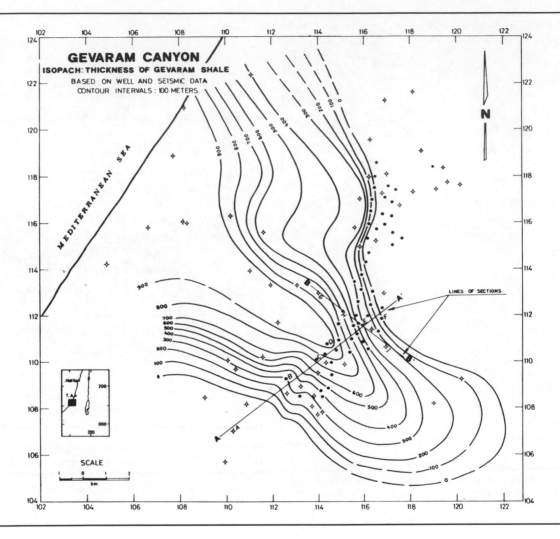

FIGURE 7.28. *Helez Oil Field, Israel, with isopach contours showing thickness of Cretaceous Gevaram Shale defining Gevaram submarine canyon (from Cohen, 1976; republished with permission of the American Association of Petroleum Geologists).*

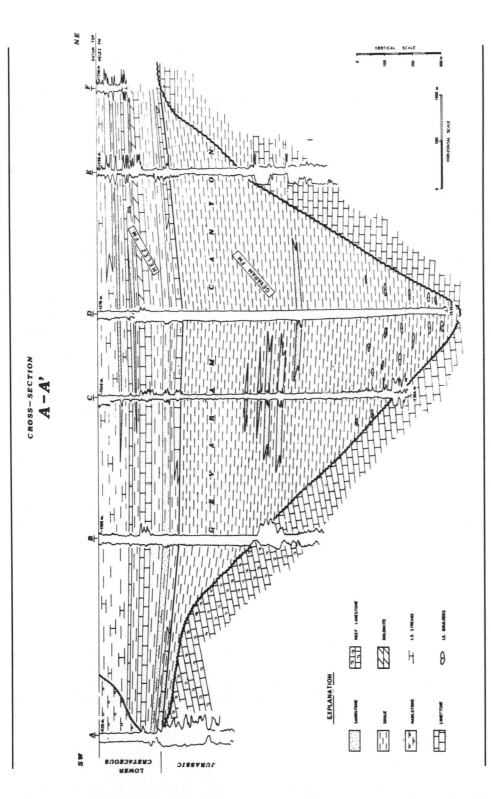

FIGURE 7.29. *Northwest-southwest cross-section A–A' (in Figure 7.28) showing shape of Gevaram submarine canyon (from Cohen, 1976; republished with permission of the American Association of Petroleum Geologists).*

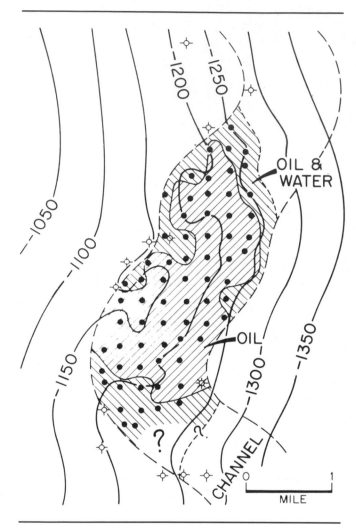

FIGURE 7.30. *Paduca Oil Field, Lea County, New Mexico, showing distribution of Permian submarine channel in Ramsey Sandstone (from Payne, 1976; republished with permission of the American Association of Petroleum Geologists).*

Rosedale, Coulter, Gosford, and Bellevue and each comprise a single cycle of deposition (MacPherson, 1978). Their distribution suggests a series of stacked submarine fan channel sand fills. Each of these sand sequences is arranged in lenticular fashion, and the SP log characteristics suggests a coarsening and thickening succession in the central portions of the channel systems, which split into smaller discrete bodies on the channel flanks (Figure 7.27). Internal growth faulting within the submarine fan complex provided additional reservoir seals.

Other examples of deep-water turbidite reservoir systems are the Hackberry Sandstone (Oligocene) of Louisiana (Paine, 1970), the Frio Sandstone (Oligocene) of the Nine Mile Point Field, Texas (R.R. Berg and Powell, 1976), the Spraberry Sands of the Midland Basin, Texas (Handford, 1981), and the Late Cretaceous sandstones of the Fazenda Cedro Field of Brazil (Alves and Beurlen, 1978). In the last three examples, a turbidite submarine fan origin was determined from preservation of Bouma sequences in cores, acoustic transparent sandstone zones in seismic sections, and the "funnel-shaped" log patterns of a coarsening-upward sequence associated with a submarine fan progradation into a basin of deposition.

Two additional examples are cited because of unusual details or uniqueness. The first is the Early Cretaceous Gevaram Shale from the Helez Oil Field of Israel. There, isopach mapping demonstrated that older limestones (Jurassic) were cut by a submarine canyon that was filled subsequently (Figures 7.28 and 7.29). Production comes from sandstone interbedded with the mudstones that filled the canyon (Figure 7.29). The log characteristics of each sandstone show a sharp base and a sloping top indicating each sand is a graded turbidite.

The second example is from the Delaware Sandstones (Permian) of the Paduca Field in the Permian Basin of West Texas and New Mexico. There, turbidite sandstones (Harms, 1974) are known to fill the basin. The Paduca Field was described by Payne (1976), who was able to map out the geometry of the submarine channel sandstone reservoir (Figure 7.30). Cores recovered from the field show preservation of complete Bouma sequences (Figure 7.31). The gamma-ray log pattern is blunt-base, blunt-top with minor fluctuations near the top. A sonic velocity log of the same interval shows an abrupt base

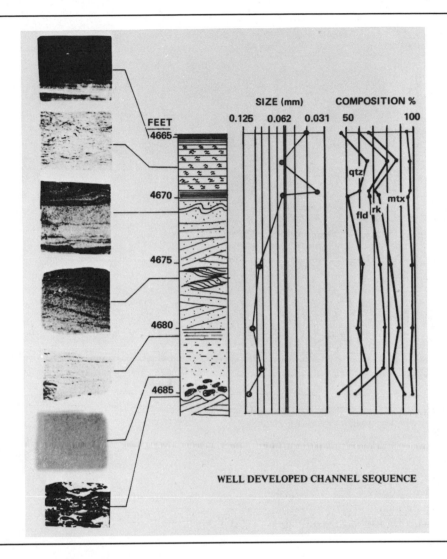

FIGURE 7.31. *Well-developed channel sequence in Ramsey Sandstone (Permian), Paduca Field, Lea County, New Mexico, showing complete components of a Bouma sequence, lithologic log, size distribution, and composition (from Payne, 1976; republished with permission of the American Association of Petroleum Geologists).*

184

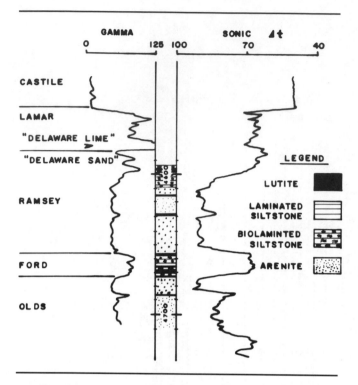

GAMMA
0   125

SONIC Δt
100   70   40

CASTILE

LAMAR

"DELAWARE LIME"

"DELAWARE SAND"

RAMSEY

FORD

OLDS

LEGEND

LUTITE

LAMINATED
SILTSTONE

BIOLAMINTED
SILTSTONE

ARENITE

FIGURE 7.32. *Gamma-ray and sonic velocity logs in Ramsey Sandstone, Paduca Field, Lea County, New Mexico (from Payne, 1976; republished with permission of the American Association of Petroleum Geologists).*

and sloping top (Figure 7.32) consistent with the fining-upward Bouma sequence (Figure 7.31) shown earlier.

In summary, the log pattern of many turbidite reservoirs reflects fining-upward of individual sandstones emplaced by turbidity currents, as part of an overall coarsening-upward and thickening-upward motif, indicating progradation of a submarine fan. In seismic section, these reservoirs show as acoustically transparent zones.

References Cited

Adams, S. S., Curtis, H. S., Hafen, P. L., and Salek-Nejad, H., 1978, Interpretation of post-depositional processes related to the formation and destruction of the Jackpile Uranium deposit, northwest New Mexico: Econ. Geol., **73**, 1635–1654.

Ahlbrandt, T. S., 1979, Textural parameters of eolian deposits: *in* McKee, E. D., 1979, A study of global sand seas: U.S. Geol. Survey Prof. Paper 1052, 21–52.

Ahlbrandt, T. S., and Fryberger, S. G., 1981, Sedimentary features and significance of interdune deposits: *in* Ethridge, F. G., and Flores, R. M., eds., Recent and ancient nonmarine depositional environments: models for exploration: Soc. Econ. Paleontologists and Mineralogists Spec. Pub. 31, 293–314.

Allen, J. R. L., 1963, Henry Clifton Sorby and the sedimentary structures of sands and sandstones in relation to flow conditions: Geol. en Mijnb., **42**, 223–228.

———, 1964, Studies of fluviatile sedimentation: six cyclothems from the Lower Old Red Sandstone, Anglo-Welsh basin: Sedimentology, **3**, 163–198.

———, 1965a, A review of the origin and characteristics of Recent alluvial sediments: Sedimentology, **5**, 88–191.

———, 1965b, Late Quaternary Niger Delta and adjacent areas: Am. Assoc. Petroleum Geologists Bull., **49**, 547–600.

———, 1970, Sediments of the modern Niger Delta: *in* Morgan, J. P., ed., Deltaic sedimentation: modern and ancient: Soc. Econ. Paleontologists and Mineralogists Spec. Pub. 15, 138–151.

———, 1982, Mud drapes in sand-wave deposits: a physical model with application to the Folkestone beds (Early Cretaceous), southeast England: Roy. Soc. London Phi. Trans., Ser. A, **306**, 291–345.

Allen, J. R. L., and Friend, P. F., 1976, Changes in intertidal dunes during two spring-neap cycles, Lifeboat Station Bank, Wells-next-the-sea, Norfolk (England): Sedimentology, **23**, 329–347.

Allen, P. A., and Homewood, P., 1984, Evolution and mechanics of a Miocene tidal sandwave: Sedimentology, **31**, 63–81.

Allen, P., and Matter, A., 1982, Oligocene meandering stream sedimentation in the eastern Ebro Basin, Spain: Eclogae Geol. Helv., **75**, 33–49.

Al-Shaieb, Z., Olmsted, R. W., Helton, J. W., May, R. T., Owens, R. T., and Hanson, R. E., 1977, Uranium potential of Permian and Pennsylvanian sandstones in Oklahoma: Am. Assoc. Petroleum Geologists Bull., **61**, 360–375.

Alves, R. J., and Beurlen, C., 1978, Model of accumulation in submarine canyon–Fazenda Cedro Field–Espirito Santo Basin: Offshore Brazil, **78**, 04.7–04.17.

Amaral, E. J., and Pryor, W. A., 1971, Texture and grain surface character of St. Peter Sandstone (Abs.): Am. Assoc. Petroleum Geologists Bull., **55**, 329–330.

Anderton, R., 1976, Tidal-shelf sedimentation: an example from the Scottish Dalradian: Sedimentology, **23**, 429–458.

Andrews, S., 1981, Sedimentology of Great Sand Dunes, Colorado: in Ethridge, F. G., and Flores, R. M., eds., Recent and ancient nonmarine depositional environments: models for exploration: Soc. Econ. Paleontologists and Mineralogists Spec. Pub. 31, 279–292.

Asaoka, O., and Moriyasu, S., 1966, On the circulation in the East China Sea and the Yellow Sea in winter: Oceanog. Mag., **18**, 1–2, 73–81.

Bagnold, R. A., 1941, The physics of blown sand and desert dunes: London, Methuen & Co., Ltd., 265 p.

——, 1954, Experiment on a gravity-free dispersion of large solid spheres in a Newtonian fluid under shear: Roy. Soc. London Proc., Ser. A., **225**, 49–63.

Barnes, J. J., and Klein, G. deV., 1975, Tidal deposits in the Zabriskie Quartzite (Cambrian), eastern California and western Nevada: in Ginsburg, R. N., ed., Tidal deposits: New York, Springer-Verlag, 163–169.

Barnes, N. E., and Normark, W. R., 1984, Diagnostic parameters for comparing modern submarine fans and ancient turbidite systems: Geo-Marine Lettr., **3**, map.

Barwis, J. H., 1978, Stratigraphy of Kiawah Island beach ridges: Southeastern Geology, **19**, 111–122.

Bates, C. C., 1953, Rational theory of delta formation: Am. Assoc. Petroleum Geologists Bull., **37**, 2119–2162.

Baumann, R. H., Day, J. W., Jr., and Miller, C. A., 1984, Mississippi Deltaic wetland survival: sedimentation versus coastal submergence: Science, **224**, 1093–1095.

Beaumont, E. A., 1979, Depositional environments of Fort Union (Tertiary, northwest Colorado) and their relation to coal: Am. Assoc. Petroleum Geologists Bull., **63**, 194–217.

Belderson, R. J., Kenyon, N. H., and Stride, A. H., 1971, Holocene sediments on the continental shelf west of the British Isles: in Delaney, F. M., ed., The geology of the east Atlantic continental margin: ICSU/SCOR Working Party 31 Symposium, Inst. Geol. Sci. Report 70, 157–170.

Belderson, R. H., Kenyon, N. H., Stride, A. H., and Stubbs, A. R., 1972, Sonographs of the sea floor: Amsterdam, Elsevier, 185 p.

Belt, E. S., 1968, Carboniferous continental sedimentation, Atlantic Provinces, Canada: in Klein, G. deV., ed., Late Paleozoic and Mesozoic continental sedimentation, northeastern North America: Geol. Soc. America Spec. Paper 106, 127–176

Benson, R. H., Sheridan, R. E., et al., 1978, Initial reports of the Deep Sea Drilling Project, v. 44: Washington, DC, U.S. Government Printing Office, 1005 p.

Berg, O. R., 1982, Seismic detection and evaluation of delta and turbidite sequences: Am. Assoc. Petroleum Geologists Bull., **66**, 1271–1288.

Berg, R. R., 1968, Point-bar origin of Fall River sandstone reservoirs, northeastern Wyoming: Am. Assoc. Petroleum Geologists Bull., **52**, 2116–2122.

——, 1975, Depositional environment of Upper Cretaceous Sussex Sandstone, House Creek Field, Wyoming: Am. Assoc. Petroleum Geologists Bull., **59**, 2099–2110.

Berg, R. R., and Davies, D. K., 1968, Origin of Lower Cretaceous Muddy Sandstone at Bell Creek Field, Montana: Am. Assoc. Petroleum Geologists Bull., **52**, 1888–1898.

Berg, R. R., and Powell, R. R., 1976, Density-flow origin for Frio Reservoir sandstones, Nine Mile Point Field, Arkansas County, Texas: Gulf Coast Assoc. Geol. Soc., **26**, 310–319.

Bernard, H. A., LeBlanc, R. J., and Major, C. F., 1962, Recent and Pleistocene geology of southeast Texas, field excursion no. 3: in Geology of the Gulf Coast and Central Texas and guidebook of excursions: Geol. Soc. America, 174–224.

Bernard, H. A., Major, C. F., Jr., Parrott, B. S., and LeBlanc, R. J., 1970, Brazos alluvial plain environment: Texas Bureau of Econ. Geol. Guidebook No. 11.

Bersier, A., 1958, Sequences detritiques et divagations fluviales: Eclogae Geol. Helv., **51**, 854–893.

Blakey, R. C., 1982, Marine sand-wave complex in the Permian of central Arizona: Jour. Sedimentary Petrology, **54**, 29–51.

Blissenbach, E., 1954, Geology of alluvial fans in semiarid regions: Geol. Soc. America Bull., **65**, 175–189.

Bloomer, R. R., 1977, Depositional environments of a reservoir sandstone in west-central Texas: Am. Assoc. Petroleum Geologists Bull., **61**, 344–359.

Bluck, B. J., 1965, The sedimentary history of some Triassic conglomerates in the Vale of Glamorgan, South Wales: Sedimentology, **4**, 225–245.

——, 1971, Sedimentation in the meandering River Endrick, Scotland: Scottish Jour. Geology, **7**, 93–138.

Boersma, J. R., and Terwindt, J. H. J., 1981, Neap-spring tide sequences of intertidal shoal deposits in a mesotidal estuary: Sedimentology, **28**, 151–170.

Boggs, S., Jr., 1974, Sand-wave fields in Taiwan Strait: Geology, **2**, 251–253.

Booth, J. S., 1979, Recent history of mass-wasting of the upper continental slopes, northern Gulf of Mexico, as interpreted from the consolidation states of the sediments: in Doyle, L. J., and Pilkey, O. H., eds., Geology of continental slopes: Soc. Econ. Paleontologists and Mineralogists Spec. Pub. 27, 153–164.

Boothroyd, J. C., 1978, Mesotidal inlets and estuaries: in Davis, R. A., Jr., ed., Coastal sedimentary environments: New York, Springer-Verlag, 287–360.

Boothroyd, J. C., and Hubbard, D. K., 1975, Genesis of bedforms in mesotidal estuaries: *in* Cronin, L. E., ed., Estuarine research, 2: New York, Academic Press, 217–234.

Bouma, A. H., 1962, Sedimentology of some flysch deposits: Amsterdam, Elsevier, 169 p.

Bouma, A. H., and Pluenneke, J. L., 1975, Structures and textural characteristics of debrites from the Philippine Sea: *in* Karig, D. E., Ingle, J. C., et al., 1975, Initial Reports of the Deep Sea Drilling Project, v. 31: U.S. Government Printing Office, 497–505.

Bouma, A. H., Stelting, C. F., and Coleman, J. M., 1984, Mississippi Fan: internal structure and depositional processes: Geo-Marine Lettr., **3**, 147–154.

Bouma, A. H., Berryhill, H. L., Knebel, H. J., and Brenner, R. L., 1982, Continental shelf: *in* Scholle, P. A., and Spearing, D., eds., Sandstone depositional environments: Am. Assoc. Petroleum Geologists Mem. 31, 281–328.

Bourgeois, J., 1980, A transgressive shelf sequence exhibiting hummocky stratification: the Cape Sebastian Sandstone (Upper Cretaceous), southwestern Oregon: Jour. Sedimentary Petrology, **50**, 681–702.

Bowen, A. J., Normark, W. R., and Piper, D. J. W., 1984, Modelling of turbidity currents on Navy submarine fan, California continental borderland: Sedimentology, **31**, 169–185.

Bowsher, A. L., 1967, Ocean tides as a geologic process: *in* Teichert, C., and Yochelson, E. L., Essays in paleontology and stratigraphy: Univ. of Kansas Dept. of Geology Spec. Pub. No. 2, 319–348.

Boyles, J. M., and Scott, A. J., 1982, A model for migrating shelf-bar sandstones in Upper Macos Shale (Campanian), northwestern Colorado: Am. Assoc. Petroleum Geologists Bull., **66**, 491–508.

Brenner, R. L., 1978, Sussex Sandstone of Wyoming—examples of Cretaceous offshore sedimentation: Am. Assoc. Petroleum Geologists Bull., **62**, 181–200.

Brenner, R. L., and Davies, D. K., 1973, Storm-generated coquinoid sandstone: genesis of high-energy marine sediments from the upper Jurassic of Wyoming and Montana: Geol. Soc. America Bull., **84**, 1685–1698.

Bridge, J. S., and Jarvis, J., 1982, The dynamics of a river bend: a study in flow and sedimentary processes: Sedimentology, **29**, 499–541.

Briggs, G., 1974, Carboniferous depositional environments in the Ouachita Mountains–Arkoma basin area of southeastern Oklahoma: *in* Briggs, G., ed., Carboniferous of the southeastern United States: Geol. Soc. America Spec. Paper 148, 225–240.

Brookfield, M. E., 1977, The origin of bounding surfaces in ancient aeolian sandstones: Sedimentology, **24**, 303–332.

Broussard, M. L., ed., 1975, Deltas, 2d ed.: Houston, Houston Geol. Soc., 555 p.

Brown, A. R., Dahm, C. G., and Graebner, R. J., 1981, A stratigraphic case history using three-dimensional seismic data in the Gulf of Thailand: Geophys. Prospecting, **29**, 327–349.

Brown, A. R., Graebner, R. J., and Dahm, C. G., 1982, Use of horizontal seismic sections to identify subtle traps: *in* Halbouty, M. T., ed., 1982, The deliberate search for the subtle trap: Am. Assoc. Petroleum Geol. Mem. 32, 47–56.

Buck, S. G., 1983, The Saaiplas Quartzite Member: a braided system of gold- and uranium-bearing channel placers within the Proterozoic Witwatersrand Supergroup of South Africa: *in* Collinson, J. D., and Lewin, J., eds., Modern and ancient fluvial systems: Int. Assoc. Sedimentologists Spec. Pub. 6, 549–562.

Bull, W. B., 1972, Recognition of alluvial fan deposits in the stratigraphic record: *in* Rigby, J. K., and Hamblin, W. K., eds., Recognition of ancient sedimentary environments: Soc. Econ. Paleontologists and Mineralogists Spec. Pub. 16, 63–83.

Burke, K., 1972, Longshore drift, submarine canyons and submarine fans in development of Niger Delta: Am. Assoc. Petroleum Geologists Bull., **56**, 1975–1983.

Busch, D. A., 1959, Prospecting for stratigraphic traps: Am. Assoc. Petroleum Geologists Bull., **43**, 2829–2843.

Busch, W. H., 1976, A model of the origin of dish structures in some ancient subaqueous sandy debris flow deposits (Abs.): Geol. Soc. America Abstracts with Programs, **8**, 799.

Button, A., and Vos, R. G., 1977, Subtidal and intertidal clastic and carbonate sedimentation in a macrotidal environment: an example from the lower Proterozoic of South Africa: Sediment. Geol., **18**, 175–200.

Campbell, C. V., 1971, Depositional model—Upper Cretaceous Gallup shoreline, Ship Rock area, New Mexico: Jour. Sedimentary Petrology, **41**, 395–409.

———, 1976, Reservoir geometry of a fluvial sheet sandstone: Am. Assoc. Petroleum Geologists Bull., **60**, 1009–1020.

Cant, D. J., 1978a, Bedforms and bar types in the South Saskatchewan River: Jour. Sedimentary Petrology, **48**, 1321–1330.

———, 1978b, Development of a facies model for sandy braided river sedimentation: comparison of the South Saskatchewan River and the Battery Point Formation: *in* Miall, A. D., ed., Fluvial sedimentology: Can. Soc. Petroleum Geol. Mem. 5, 627–639.

Cant, D. J., and Walker, R. G., 1978, Fluvial processes and facies sequences in the sandy braided South Saskatchewan River, Canada: Sedimentology, **25**, 625–648.

Carlson, P. R., and Molnia, B. F., 1977, Submarine faults and slides on the continental shelf, North Gulf of Alaska: Marine Geotech., **2**, 275–290.

Carlson, P. R., and Nelson, C. H., 1968, Sediments and sedimentary structures of the Atoria submarine canyon-fan system, northeast Pacific: Jour. Sedimentary Petrology, **39**, 1269–1282.

Caston, V. N. D., 1972, Linear sand banks in the southern North Sea: Sedimentology, **18**, 63–78.

———, 1979, The Quaternary sediments of the North Sea: *in* Banner, D. T., Collins, M. B., and Massie, K. S., eds., 1979, The northwest European shelf seas: The sea bed and the sea in motion 1. Geology and Sedimentology: Amsterdam, Elsevier, 195–270.

Chough, S. K., 1983, Marine geology of Korean seas: Boston, IHRDC, 160 p.

Chough, S. K., and Hesse, R., 1976, Submarine meandering thalweg and turbidity currents flowing for 4,000 km in the northwest Atlantic mid-ocean channel, Labrador Sea: Geology, **4**, 529–533.

———— and ————, 1980, The northwest Atlantic mid-ocean channel of the Labrador Sea. III: Head spill vs. body spill deposits from turbidity currents on natural levees: Jour. Sedimentary Petrology, **50**, 227–234.

Clark, R. H., and Rouse, J. T., 1971, A closed system for generation and entrapment of hydrocarbons in Cenozoic deltas, Louisiana Gulf Coast: Am. Assoc. Petroleum Geologists Bull., **55**, 1170–1178.

Clemmensen, L. B., and Abrahamsen, K., 1983, Aeolian stratification and facies association in desert sediments, Arran Basin (Permian), Scotland: Sedimentology, **30**, 311–339.

Clifton, H. E., Hunter, R. E., and Phillips, R. L., 1971, Depositional structures and processes in the non-barred, high energy nearshore: Jour. Sedimentary Petrology, **41**, 651–670.

Cline, L. M., 1960, Stratigraphy of the late Paleozoic rocks of the Ouachita Mountains, Oklahoma: Okla. Geol. Survey Bull., **85**, 113 p.

Cline, L. M., and Moretti, F., 1956, Two measured sections of Jackfork Group in southeastern Oklahoma: Okla. Geol. Survey Circ. 41, 20 p.

Cohen, Z., 1976, Early Cretaceous buried canyon: influence on accumulation of hydrocarbons in Helez Oil Field, Israel: Am. Assoc. Petroleum Geologists Bull., **60**, 108–114.

Coleman, J. M., 1976, Deltas: Processes of sedimentation and models for exploration: Minneapolis, CEPCO Div., Burgess Pub. Co., 102 p.

————, 1980, Deltas: Processes of sedimentation and models for exploration, 2d ed.: Boston, IHRDC, 124 p.

Coleman, J. M., and Gagliano, S. M., 1965, Sedimentary structures: Mississippi Delta Plain: in Middleton, G. V., ed., 1965, Primary sedimentary structures and their hydrodynamic significance: Soc. Econ. Paleon. and Min. Spec. Pub 12, 133–148.

Coleman, J. M., and Wright, L. D., 1975, Modern river deltas: variability of processes and sand bodies: in Broussard, M. L., ed., Deltas, 2d ed.: Houston, Houston Geol. Soc., 99–150.

Coleman, J. M., Gagliano, S. M., and Smith, W. G., 1970, Sedimentation in a Malaysian high tide tropical delta: in Morgan, J. P., ed., Deltaic sedimentation: modern and ancient: Soc. Econ. Paleontologists and Mineralogists Spec. Pub. 15, 185–197.

Collinson, J. D., and Lewin, J., eds., 1983, Modern and ancient fluvial systems: Int. Assoc. Sedimentologists Spec. Pub. 6, 575 p.

Combaz, A., and DeMatharel, M., 1978, Organic sedimentation and genesis of petroleum in Mahakam Delta, Borneo: Am. Assoc. Petroleum Geologists Bull., **62**, 1684–1695.

Cook, H. E., Jenkyns, H. C., and Kelts, K. R., 1976, Redeposited sediments along the Line Islands, Equatorial Pacific: in Schlanger, S. O., Jackson, E. D. et al., eds., Initial report of the Deep Sea Drilling Project, v. 33: Washington, DC, U.S. Government Printing Office, 837–847.

Costello, W. R., and Walker, R. G., 1972, Pleistocene sedimentology, Credit River, southern Ontario: a new component of the braided river model: Jour. Sedimentary Petrology, **42**, 389–400.

Cotter, E., 1971, Paleoflow characteristics of a Late Cretaceous river in Utah from analysis of sedimentary structures in the Ferron Sandstone: Jour. Sedimentary Petrology, **41**, 129–138.

————, 1978, The evolution of fluvial style with special reference to the central Appalachians: in Miall, A. D., ed., Fluvial sedimentology: Can. Soc. Petroleum Geol. Mem. 5, 361–384.

————, 1983, Shelf, paralic, and fluvial environments and eustatic sea-level fluctuations in the origin of the Tuscarora Formation (lower Silurian) of central Pennsylvania: Jour. Sedimentary Petrology, **53**, 25–49.

Cram, J. M., 1979, The influence of continental shelf width on tidal range: paleoceanographic implications: Jour. Geology, **87**, 441–447.

Creager, J. S., Scholl, D. W., et al., 1971, Initial reports of the Deep Sea Drilling Project, v. 19: Washington, DC, U.S. Government Printing Office.

Crowell, J. C., 1974, Sedimentation along the San Andreas Fault, California: in Dott, R. H., Jr., ed., Modern and ancient geosynclinal sedimentation: Soc. Econ. Paleontologists and Mineralogists Spec. Pub. 19, 292–303.

Curray, J. R., and Moore, D. G., 1971, Growth of the Bengal deep-sea fan and denudation of the Himalayas: Geol. Soc. America Bull., **82**, 563–572.

Curtis, D. M., and Picou, E. B., Jr., 1978, Gulf coast Cenozoic: a model for the application of stratigraphic concepts to exploration on passive margins: Gulf Coast Assoc. Geol. Soc., **28**, 103–120.

Dalrymple, R. W., Knight, R. J., and Lambiase, J. J., 1978, Bedforms and their hydrodynamic stability relationships in a tidal environment, Bay of Fundy, Canada: Nature, **275**, 100–104.

Damuth, J. E., and Embley, R. W., 1981, Mass-transport processes on Amazon Cone: western equatorial Atlantic: Am. Assoc. Petroleum Geol. Bull., **65**, 629–643.

Damuth, J. E., and Flood, R. D., 1984, Morphology, sedimentation processes and growth pattern of the Amazon deep-sea fan: Geo-Marine Lettr., **3**, 109–118.

Damuth, J. E., and Kumar, N., 1975, Amazon Cone: morphology, sediments, age and growth pattern: Geol. Soc. America Bull., **86**, 863–878.

Damuth, J. E., Kolla, V., Flood, R. D., Kowsmann, R. O., Monteiro, M. C., Gorini, M. A., Palma, J. J. C., and Belderson, R. H., 1983, Distributary channel meandering and bifurcation patterns on the Amazon Deep-Sea fan as revealed by long-range side-scan sonar (GLORIA): Geology, **11**, 94–98.

Damuth, J. E., Kowsmann, R. O., Flood, R. D., Belderson, R. H., and Gorini, M. A., 1983, Age relationships of distributary channels on Amazon deep-sea fan: implications for fan growth: Geology, **11**, 470–473.

Davies, D. K., Ethridge, F. G., and Berg, R. R., 1971, Recognition of barrier environments: Am. Assoc. Petroleum Geologists Bull., **55**, 550–565.

Davies, J. L., 1964, A morphogenetic approach to world shorelines: Zeits. Geomorphologie, **8**, 127–142.

Davis, R. A., Jr., 1985, Beach and nearshore zone: *in* Davis, R. A., Jr., ed., Coastal sedimentary environments, 2d ed.: New York, Springer-Verlag.

Davis, R. A., Jr., Fox, W. T., Hayes, M. O., and Boothroyd, J. C., 1972, Comparison of ridge and runnel systems in tidal and non-tidal environments: Jour. Sedimentary Petrol., **42**, 413–421.

DeBeaumont, E. L., 1845, Leçons de geologie practique: Paris, 7 me Leçon-levees de sables etgalets.

DeJong, J. D., 1965, Quaternary sedimentation in the Netherlands: *in* Wright, H. E., and Frey, D. G., eds., International studies on the Quaternary: Geol. Soc. America Spec. Paper 84, 95–124.

Denny, C. S., 1967, Fans and pediments: Am. Jour. Sci., **265**, 81–105.

DeRaaf, J. F. M., and Boersma, J. R., 1971, Tidal deposits and their sedimentary structures: Geol. en Mijnb., **50**, 479–504.

DeRaaf, J. F. M., Reading, H. G., and Walker, R. G., 1965, Cyclic sedimentation in the Lower Westphalian of North Devon, England: Sedimentology, **4**, 1–52.

Dickinson, K. A., 1976, Sedimentary depositional environments of uranium and petroleum host rocks of the Jackson Group, South Texas: Jour. U.S. Geol. Survey Res., **4**, 615–626.

Dickinson, K. A., and Sullivan, M. W., 1976, Geology of the Brysch uranium mine, Karnes County, Texas: Jour. U.S. Geol. Survey Res., **4**, 397–404.

Dickinson, K. A., Berryhill, H. L., Jr., and Holmes, C. W., 1972, Criteria for recognizing ancient barrier coastlines: *in* Rigby, J. K., and Hamblin, W. K., eds., Recognition of ancient sedimentary environments: Soc. Econ. Paleontologists and Mineralogists Spec. Pub. 16, 192–214.

Dietz, R. S., 1963, Wave base, marine profile of equilibrium and wave-built terraces: a critical appraisal: Geol. Soc. America Bull., **74**, 971–990.

Dill, R. F., 1964, Sedimentation and erosion in Scripps Submarine Canyon head: *in* Miller, R. L., ed., Papers in marine geology: New York, Macmillan Publishing Co., 23–41.

Doeglas, D. J., 1962, The structure of sedimentary deposits in braided rivers: Sedimentology, **1**, 167–190.

Dott, R. H., Jr., and Bourgeois, J., 1982, Hummocky stratification: significance of its variable bedding sequences: Geol. Soc. America Bull., **93**, 663–680.

Doyle, L. J., Pilkey, O. H., and Woo, C. C., 1979, Sedimentation on the eastern United States continental slope: *in* Doyle, L. E., and Pilkey, O. H., eds., Geology of continental slopes: Soc. Econ. Paleontologists and Mineralogists Spec. Pub. 27, 119–130.

Drake, D. E., Cacchione, D. A., Muensch, R. D., and Nelson, C. H., 1980, Sediment transport in Norton Sound, Alaska: Marine Geol., **36**, 97–126.

Driese, S. G., Byers, C. W., and Dott, R. H., Jr., 1981, Tidal deposition in the basal Upper Cambrian Mt. Simon Formation in Wisconsin: Jour. Sedimentary Petrology, **51**, 367–381.

Duane, D. B., Field, M. E., Meisburger, E. P., Swift, D. J. P., and Williams, S. J., 1972, Linear shoals on the Atlantic innercontinental shelf, Florida to Long Island: *in* Swift, D. J. P., Duane, D. B., and Pilkey, O. H., 1972, Shelf sediment transport: process and pattern: Stroudburg, PA, Dowden, Hutchinson & Ross, Inc., 447–498.

Dutton, S. P., 1982, Pennsylvanian Fan-Delta and Carbonate deposition, Mobeetie Field, Texas Panhandle: Am. Assoc. Petroleum Geologists Bull., **66**, 389–407.

Dzulynski, S., Ksiaskiewicz, M., and Kuenen, P. H., 1959, Turbidites in flysch of the Polish Carpathian Mountains: Geol. Soc. America Bull., **70**, 1089–1118.

Eargle, D. H., Dickinson, K. A., and Davis, B. A., 1975, South Texas uranium deposits: Am. Assoc. Petroleum Geologists Bull., **59**, 766–779.

Edwards, M. B., 1981, Upper Wilcox Rosita delta system of south Texas: growth-faulted shelf-edge deltas: Am. Assoc. Petroleum Geologists Bull., **65**, 54–73.

Elmore, R. D., 1984, The Copper Harbor Conglomerate: a late Precambrian fining-upward alluvial fan sequence in northern Michigan: Geol. Soc. America Bull., **95**, 610–617.

Embley, R. W., 1976, New evidence for occurrence of debris flow deposits in the deep sea: Geology, **4**, 371–374.

Emmel, F. J., and Curray, J. R., 1984, The Bengal submarine fan, northeastern Indian Ocean: Geo-Marine Lettr., **3**, 119–124.

Emery, K. O., 1968, Relict sediments on continental shelves of the world: Am. Assoc. Petroleum Geologists Bull., **52**, 445–464.

———, 1980, Continental margins—classification and petroleum prospects: Am. Assoc. Petroleum Geologists Bull., **64**, 297–315.

Emery, K. O., and Milliman, J. D., 1978, Suspended matter in surface waters: influence of river discharge and upwelling: Sedimentology, **25**, 125–140.

Emery, K. O., Hayashi, Y., Hilde, T. W. C., Kobayashi, K., Koo, J. H., Meng, C. Y., Niino, H., Osterhagen, J. H., Reynolds, L. M., Wageman, J. M., Wang, C. S., and Yang, S. J., 1969, Geological structure and some water characteristics of the East China Sea and the Yellow Sea: Tech. Bull., ECAFE, **2**, 3–43.

Emery, K. O., Wigley, R. L., Bartlett, A. S., Rubin, M., and Barghoorn, E. S., 1967, Freshwater peat on the continental shelf: Science, **158**, 1301–1307.

190

Enos, P., 1969, Anatomy of a flysch: Jour. Sedimentary Petrology, **39,** 680–723.

Eriksson, K. A., 1977, Tidal deposits from the Archean Moodies Group, Barberton Mountain Land, South Africa: Sediment. Geol., **18,** 223–264.

————, 1979, Marginal marine depositional processes from the Archean Moodies Group, Barberton Mountain Land, South Africa: evidence and significance: Precambrian Res., **8,** 153–182.

Eriksson, K. A., and Vos, R. G., 1979, A fluvial fan depositional model for Middle Proterozoic red beds from the Waterberg Group, South Africa: Precambrian Res., **9,** 169–188.

Ethridge, F. G., Jackson, T. J., and Youngberg, A. D., 1981, Flood-basin sequence of a fine-grained meander belt subsystem: the coal-bearing lower Wasatch and upper Fort Union formations, southern Powder River Basin, Wyoming: in Ethridge, F. G., and Flores, R. M., eds., Recent and ancient nonmarine depositional environments: models for exploration: Soc. Econ. Paleontologists and Mineralogists Spec. Pub. 31, 191–212.

Evans, G., 1965, Intertidal flat sediments and their environments of deposition in The Wash: Geol. Soc. London Quar. Jour., **121,** 209–241.

Evans, W. E., 1970, Imbricate linear sandstone bodies of Viking Formation in Dodsland–Hoosier area of southwestern Saskatchewan: Am. Assoc. Petroleum Geologists Bull., **54,** 469–486.

Ferm, J. C., 1962, Petrology of some Pennsylvanian sedimentary rocks: Jour. Sedimentary Petrology, **32,** 104–123.

————, 1970, Allegheny deltaic deposits: in Morgan, J. P., ed., Deltaic sedimentation: modern and ancient: Soc. Econ. Paleontologists and Mineralogists Spec. Pub. 15, 246–255.

————, 1974, Carboniferous environmental models in eastern United States and their significance: in Briggs, G., ed., Carboniferous of the southeastern United States: Geol. Soc. America Spec. Pub. 148, 79–95.

Ferm, J. C., and Cavaroc, V. V., Jr., 1968, A nonmarine sedimentary model for the Allegheny rocks of West Virginia: in Klein, G. deV., ed., Late Paleozoic and Mesozoic continental sedimentation, northeastern North America: Geol. Soc. America Spec. Paper 106, 1–20.

Ferm, J. C., and Williams, E. G., 1963, Model for Cyclic sedimentation in the Appalachian Pennsylvanian (Abs.): Amer. Assoc. Petroleum Geologists Bull., **47,** 356.

———— and ————, 1964, Sedimentary facies in the lower Allegheny rocks of western Pennsylvania: Jour. Sedimentary Petrology, **34,** 610–614.

Field, M. E., 1980, Sand bodies on coastal plain shelves: Holocene record of the U.S. Atlantic inner shelf off Maryland: Jour. Sedimentary Petrology, **50,** 505–528.

Field, M. E., and Duane, D. B., 1976, Post-Pleistocene history of the United States inner continental shelf: significance to origin of barrier islands: Geol. Soc. America Bull., **87,** 691–702.

Field, M. E., Meisburger, E. D., Stanley, E. A., and Williams, S. J., 1979, Upper Quaternary peat deposits on the Atlantic inner shelf of the United States: Geol. Soc. America Bull., **90,** 618–628.

Finley, R. J., 1978, Ebb-tidal delta morphology and sediment supply in relation to season wave energy flux, North Inlet, South Carolina: Jour. Sedimentary Petrology, **8,** 227–238.

Fischer, A. G., 1961, Stratigraphic record of transgressing seas in light of sedimentation on Atlantic coast of New Jersey: Am. Assoc. Petroleum Geologists Bull., **45,** 1656–1667.

Fischer, R. P., 1970, Similarities, differences and some genetic problems of the Wyoming and Colorado Plateau types of uranium deposits in sandstones: Econ. Geol., **65,** 778–784.

————, 1974, Exploration guides to new uranium districts and belts: Econ. Geol., **69,** 362–376.

Fisher, J. J., 1968, Barrier island formation: discussion: Geol. Soc. America Bull., **79,** 1421–1426.

Fisher, W. L., and McGowen, J. H., 1969, Depositional systems in Wilcox Group (Eocene) of Texas and their relation to occurrence of oil and gas: Am. Assoc. Petroleum Geologists Bull., **53,** 30–54.

Fisher, W. L., Galloway, W. E., Proctor, C. V., Jr., and Nagle, J. S., 1971, Depositional systems in the Jackson Group of Texas: Gulf Coast Assoc. Geol. Soc. Trans., **20,** 234–261.

Fisk, H. N., 1944, Geological investigation of the alluvial valley of the Lower Mississippi River: Vicksburg, Mississippi River Commission, 78 p.

————, 1947, Fine-grained alluvial deposits and their effect on Mississippi River activity: Vicksburg, Mississippi River Commission, 82 p.

————, 1956, Nearshore sediments of the Continental Shelf off Louisiana: Proc. Eight Texas Conf. on Soil Mechanics and Foundation Engineering, 1–23.

————, 1961, Bar-finger sands of the Mississippi Delta: in Peterson, J. A., and Osmund, J. C., eds., Geometry of sandstone bodies: Am. Assoc. Petroleum Geologists, 29–52.

Fisk, H. N., McFarlan, E., Jr., Kolb, C. R., and Wilbert, L. J., Jr., 1954, Sedimentary framework of the modern Mississippi Delta: Jour. Sedimentary Petrology, **24,** 76–99.

Fitzgerald, D. M., 1982, Sediment bypassing at mixed energy tidal inlets: Eighth Coastal Engineering Conf. Proc., 1094–1118.

Fitzgerald, D. M., and Nummedal, D., 1983, Response characteristics of an ebb-dominated tidal inlet channel: Jour. Sedimentary Petrology, **53,** 833–845.

Fitzgerald, D. M., Fink, K. R., Jr., and Lincoln, J. M., 1984, A flood-dominated mesotidal inlet: Geo-Marine Lettr., **3,** 17–22.

Flores, R. M., 1981, Coal deposition in fluvial paleoenvironments of the Paleocene Tongue River Member of the Fort Union Formation, Powder River area, Powder River Basin, Wyoming: in Ethridge, F. G., and Flores, R. M., eds., Recent and ancient nonmarine depositional environments: models for exploration: Soc. Econ. Paleontologists and Mineralogists

Spec. Pub. 31, 169–190.

Flores, R. M., Blanchard, L. F., Sanchez, J. D., Marley, W. E., and Muldoon, W. J., 1984, Paleogeographic controls of coal accumulation, Cretaceous Blackhawk Formation and Star Point Sandstone, Wasatch Plateau, Utah: Geol. Soc. America Bull., **95**, 540–550.

Folk, R. L., 1968, Bimodal supermature sandstones—produce of the desert floor: XXIII Int. Geol. Cong. Proc., **8**, 9–32.

———, 1978, Angularity and silica coatings of Simpson desert sand grains, Northern Territory, Australia: Jour. Sedimentary Petrology, **48**, 611–624.

Foristall, G. Z., 1974, Three-dimensional structure of storm-generated currents: Jour. Geophys. Res., **79**, 2721–2729.

Fox, W. T., and Davis, R. J., Jr., 1978, Seasonal variation in beach erosion and sedimentation on the Oregon Coast: Geol. Soc. America Bull., **89**, 1541–1549.

Friedkin, J. F., 1945, A laboratory study of the meandering of the alluvial rivers: Vicksburg, Mississippi River Commission, 40 p.

Friend, P. F., 1978, Distinctive features of some ancient river systems: in Miall, A. D., ed., Fluvial sedimentology: Can. Soc. Petroleum Geol. Mem. 5, 531–542.

Fryberger, S. C., 1979a, Dune forms and wind regimes: in McKee, E. D., ed., 1979, A study of global sand seas: U.S. Geol. Survey Prof. Paper 1052, 137–170.

———, 1979b, Eolian fluvial (continental) origin of ancient stratigraphic traps for petroleum in Weber Sandstone, Rangeley Oil Field, Colorado: Mountain Geologist, **16**, 1–36.

Fryberger, S. C., and Schenk, C., 1981, Wind sedimentation tunnel experiments on the origin of eolian strata: Sedimentology, **28**, 805–821.

Fryberger, S. C., Al-Sar, A. M., and Clisham, T. J., 1983, Eolian dune, interdune, sand sheet and siliciclastic sabkha sediments of an offshore prograding sand sea, Dharan Area, Saudi Arabia: Am. Assoc. Petroleum Geologists Bull., **67**, 280–312.

Gadd, P. E., Lavelle, J. W., and Swift, D. J. P., 1978, Estimates of sand transport on the New York Shelf using near-bottom current meter observations: Jour. Sedimentary Petrology, **48**, 239–252.

Galloway, W. E., 1975, Process framework for describing the morphological and stratigraphic evolution of deltaic depositional systems: in Broussard, M. L., ed., Deltas, 2d ed.: Houston, Houston Geol. Soc., 87–98.

———, 1976, Sediments and stratigraphic framework of the Copper River Fan-delta, Alaska: Jour. Sedimentary Petrology, **46**, 726–737.

———, 1977, Catahoula Formation of the Texas Coastal Plain: Depositional systems, composition, structural development, ground water flow history, and uranium distribution: Texas Bureau of Econ. Geol., Rept. of Inv. 87, 59 p.

Galloway, W. E., Hobday, D. K., and Magara, K., 1982, Frio Formation of Texas Gulf Coastal Plain: depositional systems, structural framework, and hydrocarbon distribution: Am. Assoc. Petroleum Geologists Bull., **66**, 649–688.

Garcia, R., 1981, Depositional systems and their relation to gas accumulations in Sacramento Valley, California: Am. Assoc. Petroleum Geologists Bull., **65**, 653–673.

Gardner, J. V., Field, M. E., Masson, D. G., and Parson, L. M., 1984, Surface morphology of upper and mid Laurentian Fan as seen on GLORIA long-range side-scan (Abs.): Soc. Econ. Paleontologists and Mineralogists Abstracts, Annual Midyear Meeting, 34.

Garrison, L. E., Kenyon, N. H., and Bouma, A. H., 1982, Channel systems and lobe construction in the Mississippi Fan: Geo-marine Lettr., **2**, 31–39.

Gellatly, D. C., 1970, Cross-bedded tidal megaripples from King Sound: Sediment. Geol., **4**, 185–192.

Gilbert, G. K., 1885, The topographic features of lakes shores: U.S. Geol. Survey Fifth Ann. Rept., 69–123.

Ginsburg, R. N., ed., 1975, Tidal deposits: New York, Springer-Verlag, 428 p.

Glennie, K. W., 1972, Permian Rotliegendes of northwest Europe interpreted in light of modern desert sedimentation studies: Am. Assoc. Petroleum Geologists Bull., **56**, 1048–1071.

Gould, H. R., 1970, The Mississippi Delta Complex: in Morgan, J. P., ed., Deltaic sedimentation: modern and ancient: Soc. Econ. Paleontologists and Mineralogists Spec. Pub. 15, 3–30.

Graham, S. A., and Bachman, S. B., 1983, Structural controls on submarine-fan geometry and internal architecture: upper LaJolla Fan system, offshore southern California: Am. Assoc. Petroleum Geologists Bull., **67**, 83–96.

Hageman, B. P., 1972, Sedimentation in the lowest part of river systems in relation to the post-glacial sea level rise in the Netherlands: XXIV Int. Geol. Cong. Reports, **12**, 37–47.

Hamblin, A. P., and Walker, R. G., 1979, Storm-dominated shallow marine deposits: the Fernie-Kootenay (Jurassic) transition, southern Rocky Mountains: Can. Jour. Earth Sci., **16**, 1673–1690.

Hampton, M. A., 1972, The role of subaqueous debris flow in generating turbidity currents: Jour. Sedimentary Petrology, **32**, 775–793.,

———, 1979, Buoyancy in debris flows: Jour. Sedimentary Petrology, **49**, 753–758.

Hampton, M. A., and Bouma, A. H., 1977, Shelf instability near the shelf break, western Gulf of Alaska: Marine Geotech., **2**, 309–331.

Hand, B. M., Hayes, M. O., and Wessell, J. M., 1969, Antidunes in the Mount Toby Formation (Triassic), Massachusetts: Jour. Sedimentary Petrology, **39**, 1310–1316.

Handford, C. R., 1981, Sedimentology and genetic stratigraphy of Dean and Spraberry formations (Permian), Midland Basin, Texas: Am. Assoc. Petroleum Geologists Bull., **65**, 1602–1616.

Haner, B. E., 1971, Morphology and sediments of Redondo submarine fan, southern California: Geol. Soc. America Bull., **82**, 2413–2431.

Hantzschel, W., 1939, Tidal flat deposits: *in* Trask, P. D., ed., Recent marine sediments: Soc. Econ. Paleontologists and Mineralogists Spec. Pub. 1, 195–200.

Harms, J. C., 1966, Stratigraphic traps in a valley fill, western Nebraska: Am. Assoc. Petroleum Geologists Bull., **50,** 2119–2149.

———, 1974, Brushy Canyon Formation, Texas: a deep-water density current deposit: Geol. Soc. America Bull., **85,** 1763–1784.

Harms, J. C., and Tachenberg, P., 1972, Seismic signatures of sedimentation models: Geophysics, **37,** 45–58.

Harms, J. C., MacKenzie, D. B., and McCubbin, D. G., 1963, Stratification in modern sands of the Red River, Louisiana: Jour. Geology, **71,** 556–580.

Harms, J. C., Southard, J. B., Spearing, D. R., and Walker, R. G., 1975, Depositional environments as interpreted from primary sedimentary structures and stratification sequences: Soc. Econ. Paleontologists and Mineralogists Short Course Notes 2, 161 p.

Hayes, M. O., 1967, Hurricanes as geologic agents: case studies of hurricanes Carla, 1961, and Cindy, 1963: Texas Bureau of Econ. Geol. Rept. of Inv., **61,** 1–54.

———, ed., 1969, Coastal environments: Northeast Massachusetts and New Hampshire: Eastern Section, Soc. Econ. Paleontologists and Mineralogists Guidebook, 462 p.

———, 1975, Morphology of sand accumulation in estuaries: *in* Cronin, L. E., ed., Estuarine research, 2: New York, Academic Press, 3–22.

———, 1979, Barrier island morphology as a function of tidal and wave regime: *in* Leatherman, S. P., ed., Barrier islands: New York, Academic Press, 1–27.

Hayes, M. O., and Kana, T. W., 1976, Terrigenous clastic depositional environments, some modern examples: Univ. of South Carolina Tech. Report 11-CRD, 131 p.

Heezen, B. C., 1963, Turbidity currents: *in* Hill, M. N., ed., The sea, 3: New York, Wiley, 742–775.

Heezen, B. C., and Ewing, M., 1952, Turbidity currents and submarine slumps, and the 1929 Grand earthquake: Am. Jour. Sci., **250,** 849–873.

Hein, J. R., 1973, Increasing rate of movement with time between California and the Pacific Plate: from Delgada submarine fan sparce areas: Jour. Geophys. Res., **78,** 7752–7762.

Heritier, F. E., Lossel, P., and Wathne, E., 1979, Frigg Field—large submarine-fan trap in Lower Eocene rocks of North Sea Viking Graben: Am. Assoc. Petroleum Geologists Bull., **63,** 1999–2020.

Heward, A. P., 1978a, Alluvial fan and lacustrine sediments from the Stephanian A and B (LaMagdalena, Cinera-Matallana and Sabero) coalfields, northern Spain: Sedimentology, **25,** 451–488.

———, 1978b, Alluvial fan sequence and megasequence models with examples from Westphalian D-Stephanian B Coalfields, northern Spain: *in* Miall, A. D., ed., 1978, Fluvial sedimentology: Can. Soc. Petroleum Geol. Mem. 5, 669–702.

Hine, A. C., 1979, Mechanisms of berm development and resulting beach growth along a barrier spit complex: Sedimentology, **26,** 333–351.

Hobday, D. K., 1974, Beach- and barrier-island facies in the Upper Carboniferous of northern Alabama: *in* Briggs, G., ed., Carboniferous of the southeastern United States: Geol. Soc. America Spec. Paper 148, 209–224.

———, 1977, Late Quaternary sedimentary history of Inhaca Island, Mozambique: Geol. Soc. South Africa Trans., **80,** 183–191.

———, 1978, Fluvial deposits of the Ecca and Beaufort Groups in the eastern Karoo Basin, south Africa: *in* Miall, A. D., ed., Fluvial sedimentology: Can. Soc. Petroleum Geol. Mem. 5, 413–431.

Hobday, D. K., and Matthew, D., 1975, Late Paleozoic fluviatile and deltaic deposits in the northeast Karoo Basin, South Africa: *in* Broussard, M. L., ed., Deltas, models for exploration, 2d ed.: Houston, Houston Geol. Soc., 457–469.

Hobson, J. P., Jr., Fowler, M. L., and Beaumont, E. A., 1982, Depositional and statistical exploration models, upper Cretaceous offshore sandstone complex, Sussex Member, House Creek Field, Wyoming: Am. Assoc. Petroleum Geol. Bull., **66,** 697–707.

Homewood, P., and Allen, P., 1981, Wave-, tide-d and current-controlled sandbodies of Miocene molasse, western Switzerland: Am. Assoc. Petroleum Geologists Bull., **65,** 2534–2545.

Hooke, J. M., and Harvey, A. M., 1983, Meander changes in relation to bend morphology and secondary flows: *in* Collinson, J. D., and Lewin, J., eds., Modern and ancient fluvial systems: Int. Assoc. Sedimentologists Spec. Pub. 6, 121–132.

Hooke, R. L., 1967, Processes on arid-region alluvial fans: Jour. Geology, **75,** 438–460.

Horne, J. C., Ferm, J. C., Caruccio, F. T., and Baganz, B. P., 1978, Depositional models in coal exploration and mine planning in Appalachian region: Am. Assoc. Petroleum Geologists Bull., **62,** 2379–2411.

Houbolt, J. J. H. C., 1968, Recent sediments in the southern bight of the North Sea: Geol. en Mijnb., **47,** 245–273.

Houbolt, J. J. H. C., and Jonkers, J. B. M., 1968, Recent sediments in the eastern part of Lake Geneva: Geol. en Mijnb., **47,** 131–148.

Howell, D. G., and Vedder, J. G., 1984, Ferrelo Fan, California: depositional systems influenced by eustatic sea level changes: Geo-Marine Lettr., **3,** 187–192.

Hoyt, J. H., 1967, Barrier island formation: Geol. Soc. America Bull., **78,** 1125–1136.

Hsu, K. J., 1977, Studies of Ventura Field, California, 1: Facies geometry and genesis of lower Pliocene turbidites: Am. Assoc. Petroleum Geologists Bull., **61,** 137–168.

Huang, T. C., and Goodell, H. G., 1970, Sediments and sedimentary processes of eastern Mississippi Cone, Gulf of Mexico: Am. Assoc. Petroleum Geologists Bull., **54,** 2070–2100.

Hubbard, D. K., Oertel, G. F., and Nummedal, D., 1979, The role of waves and tidal currents in the development of tidal-inlet sedimentary structures and sand body geometry: examples from North Carolina, South Carolina and Georgia: Jour. Sedimentary Petrology, **49,** 1073–1092.

Hubert, J. F., and Mertz, K. A., 1980, Eolian dune field of the Triassic age, Fundy Basin, Nova Scotia: Geology, **8,** 516–519.

Hubert, J. F., Butera, J. G., and Rice, R. F., 1972, Sedimentology of the Upper Cretaceous Cody-Parkman Delta, southwestern Powder River Basin, Wyoming: Geol. Soc. America Bull., **83,** 1649–1670.

Hunter, R. E., 1977a, Basic types of stratification in small eolian dunes: Sedimentology, **24,** 361–387.

———, 1977b, Terminology of cross-stratified sedimentary layers and climbing-ripple structures: Jour. Sedimentary Petrology, **47,** 697–706.

———, 1981, Stratification styles in eolian sandstones: some Pennsylvanian to Jurassic examples from the Western Interior, U.S.A.: *in* Ethridge, F. G., and Flores, R. M., eds., Recent and ancient nonmarine depositional environments: models for exploration: Soc. Econ. Paleontologists and Mineralogists Spec. Pub. 31, 315–330.

Jackson, R. G., II, 1975, Velocity-bedform-texture patterns of meander bends in the lower Wabash River of Illinois and Indiana: Geol. Soc. America Bull., **86,** 1511–1522.

———, 1976, Depositional model of point bars in the lower Wabash River: Jour. Sedimentary Petrology, **46,** 579–594.

———, 1978, Preliminary evaluation of lithofacies models for meandering alluvial streams: *in* Miall, A. D., ed., Fluvial sedimentology: Can. Soc. Petroleum Geologists Mem. 5, 543–576.

Johnson, D. W., 1919, Shore processes and shoreline development: New York, Wiley, 584 p.

Johnson, H. D., 1977, Shallow marine sand bar sequences: an example from the late Precambrian of North Norway: Sedimentology, **24,** 245–270.

Joplin, A. V., 1966, Some applications of theory and experiment to the study of bedding genesis: Sedimentology, **7,** 71–102.

Jordan, G. F., 1962, Large submarine and sand waves: Science, **136,** 839–848.

Kastens, K. A., 1984, Earthquakes as a triggering mechanism for debris flows and turbidites on the Calabrian Ridge: Marine Geol., **55,** 13–33.

Kastens, K. A., and Cita, M., 1981, Tsunami-induced sediment transport in the abyssal Mediterranean Sea: Geol. Soc. America Bull., **82,** 845–857.

Keller, G. H., Lambert, D. W., and Bennett, R. H., 1979, Geotechnical properties of continental slope deposits—Cape Hatteras to Hydrographer Canyon: *in* Doyle, L. E., and Pilkey, O. H., eds., Geology of continental slopes: Soc. Econ. Paleontologists and Mineralogists Spec. Pub. 27, 131–153.

Keller, G. H., Lambert, D., Rowe, G., and Staresinic, N., 1973, Bottom currents in the Hudson Canyon: Science, **180,** 181–183.

Kellerhals, P., and Murray, J. W., 1969, Tidal flats at Boundary Bay, Fraser River Delta, British Columbia: Can. Soc. Petroleum Geol. Bull., **17,** 67–91.

Kelts, K., and Arthur, M. A., 1981, Turbidites after ten years of deep-sea drilling—wringing out the mop?: *in* Warme, J. E., Douglas, R. G., and Winterer, E. L., eds., The Deep Sea Drilling Project: a decade of progress: Soc. Econ. Paleontologists and Mineralogists, Spec. Pub. 32, 91–128.

Kenyon, N. H., Belderson, R. H., Stride, A. H., and Johnson, M. A., 1981, Offshore tidal sand-banks as indicators of net sand transport and as potential deposits: *in* Nio, S. D., Shuttenhelm, R. T. E., and Van Weering, T.C.E., eds., Holocene marine sedimentation in the North Sea Basin: Int. Assoc. Sedimentologists Spec. Pub. 5, 257–268.

Kidd, R. B., and Roberts, D. G., 1982, Long-range side-scan sonar studies of large-scale sedimentary features in the North Atlantic: Geol. Bassin d'Aquitaine Bull., **31,** 11–29.

Klein, G. deV., 1962, Triassic sedimentation, Maritime Provinces, Canada: Geol. Soc. America Bull., **73,** 1127–1146.

———, 1963, Bay of Fundy intertidal zone sediments: Jour. Sedimentary Petrology, **33,** 844–854.

———, 1966, Dispersal and petrology of the sandstones of the Stanley–Jackfork boundary, Ouachita Fold Belt, Arkansas and Oklahoma: Am. Assoc. Petroleum Geologists Bull., **50,** 308–326.

———, 1970a, Depositional and dispersal dynamics of intertidal sand bars: Jour. Sedimentary Petrology, **40,** 1095–1127.

———, 1970b, Tidal origin of a Precambrian quartzite—The Lower fine-grained Quartzite (Dalradian) of Islay, Scotland: Jour. Sedimentary Petrology, **40,** 973–985.

———, 1971, A sedimentary model for determining paleotidal range: Geol. Soc. America Bull., **82,** 2585–2592.

———, 1972a, Sedimentary model for determining paleotidal range: reply: Geol. Soc. America Bull., **82,** 539–546.

———, 1972b, Determination of paleotidal range in clastic sedimentary rocks: XXIV Int. Geol. Congress Rept. 6, 397–405.

———, 1974, Estimating water depths from analysis of barrier island and deltaic sedimentary sequences: Geology, **2,** 409–412.

———, 1975a, Paleotidal range sequences, Middle Member, Wood Canyon Formation (Late Precambrian), eastern California and western Nevada: *in* Ginsburg, R. N., ed., Tidal deposits: New York, Springer-Verlag, 171–177.

———, 1975b, Tidalites in the Eureka Quartzite (Ordovician), eastern California and western Nevada: *in* Ginsburg, R. N., ed., Tidal deposits: New York, Springer-Verlag, 145–151.

———, 1975c, Sedimentary tectonics in the southwest Pacific marginal basins based on Leg 30 Deep Sea Drilling Project cores from the South Fiji, Hebrides and Coral Sea Basin: Geol. Soc. America Bull., **86,** 1012–1018.

———, 1976, Holocene tidal sedimentation: Stroudsburg, Pennsylvania, Dowden, Hutchinson and Ross, 423 p.

———, 1977a, Clastic tidal facies: Minneapolis, CEPCO Div., Burgess Pub. Co., 149 p.

———, 1977b, Tidal circulation model for deposition of clastic sediments in epeiric and mioclinal shelf seas: Sediment. Geol., **18,** 1–12.

————, 1980, Sandstone depositional models for exploration for fossil fuels, 2d ed.: Minneapolis, Burgess Pub. Co., 149 p.

————, 1982, Probable sequential arrangement of depositional systems on cratons: Geology, **10**, 17–22.

————, 1984, Relative rates of tectonic uplift as determined from episodic turbidite deposition in marine basins: Geology, **12**, 48–50.

————, 1985a, Intertidal flats and intertidal sand bodies: in Davis, R. A., Jr., ed., Coastal sedimentary environments, 2d ed.: New York, Springer-Verlag, 187–224.

————, 1985b, The control of depositional depth, tectonic uplift and volcanism on sedimentation processes in the back-arc basins of the western Pacific: Jour. Geology, **93**, 1–25.

————, In press, The frequency and periodicity of preserved turbidites in submarine fans as a quantitative record of tectonic uplift in collision zones: Tectonophysics.

Klein, G. deV., and Ryer, T. A., 1978, Tidal circulation patterns in Precambrian Paleozoic and Cretaceous epeiric and mioclinal shelf seas: Geol. Soc. America Bull., **89**, 1050–1058.

Klein, G. deV., DeMelo, U., and Della Favera, J. C., 1972, Subaqueous gravity processes on the front of Cretaceous deltas, Roconcavo Basin, Brazil: Geol. Soc. America Bull., **83**, 1469–1492.

Klein, G. deV., Park, Y. A., Chang, J. H., and Kim, C. S., 1982, Sedimentology of a subtidal, tide-dominated sand body in the Yellow Sea, southwest Korea: Marine Geol., **50**, 221–240.

Knight, R. J., and Dalrymple, R. W., 1975, Intertidal sediments from the south shore of Cobequid Bay, Bay of Fundy, Nova Scotia: in Ginsburg, R. N., ed., Tidal deposits: New York, Springer-Verlag, 47–56.

Kocurek, G., 1981, Significance of interdune deposits and bounding surfaces in aeolian dunes: Sedimentology, **28**, 753–780.

Kocurek, G., and Dott, R. H., Jr., 1981, Distinctions and uses of stratification types in the interpretation of eolian sand: Jour. Sedimentary Petrology, **51**, 579–595.

Kocurek, G., and Fielder, G., 1982, Adhesion structures: Jour. Sedimentary Petrology, **52**, 1229–1241.

Kolb, C. R., and Van Lopik, J. R., 1966, Depositional environments of the Mississippi River deltaic plain, southeastern Louisiana: in Broussard, M. L., and Ragsdale, J. A., eds., Deltas: Houston, Houston Geol. Soc., 17–62.

Kolla, V., and Buffler, R. T., 1984, Morphologic, acoustic and sedimentologic characteristics of the Magdalena Fan: Geo-Marine Lettr., **3**, 85–92.

Kolla, V., and Coumes, F., 1984, Morpho-acoustic and sedimentologic characteristics of the Indus Fan: Geo-Marine Lettr., **3**, 133–140.

Kolla, V., Buffler, R. T., and Ladd, J. W., 1984, Seismic stratigraphy and sedimentation of Magdalena Fan, south Columbian Basin, Caribbean Sea: Am. Assoc. Petroleum Geologists Bull., **68**, 316–332.

Komar, P. D., 1969, The channelized flow of turbidity currents with application to Monterey deep-sea fan channel: Jour. Geophys. Res., **74**, 4544–4558.

————, 1972, Relative significance of head and body spill from a channelized turbidity current: Geol. Soc. America Bull., **83**, 1151–1156.

————, 1976, Beach processes and sedimentation: Englewood Cliffs, New Jersey, Prentice-Hall, 429 p.

————, 1983, Shapes of streamlined islands on Earth and Mars: experiments and analyses of the minimum drag form: Geology, **11**, 651–654.

————, 1984, The Lemniscate loop—comparisons with the shapes of streamlined landforms: Jour. Geology, **92**, 133–145.

Kraft, J. C., 1971, Sedimentary environment facies patterns and geologic history of a Holocene transgression: Geol. Soc. America Bull., **82**, 2131–2158.

————, 1978, Coastal stratigraphic sequences: in Davis, R. A., Jr., ed., Coastal sedimentary environments: New York, Springer-Verlag, 361–384.

Kraft, J. C., and John, C. J., 1979, Lateral and vertical facies relations of transgressive barriers: Am. Assoc. Petroleum Geologists Bull., **63**, 2145–2163.

Kreisa, R. D., 1981, Storm-generated sedimentary structures in subtidal marine facies with examples from the middle and upper Ordovician of southwestern Virginia: Jour. Sedimentary Petrology, **51**, 823–848.

Kruit, C., Brouwer, J., Knox, H., Schollenberger, W., and Van Vliet, A., 1975, Une excursion aux cones d'alluvions en eau profonde d'age Tertiare pres de San Sebastian: Guidebook, Excursion Z-23, IX Int. Sed. Cong., Nice.

Kuenen, P. H., 1960, Experimental abrasion 4: Eolian action: Jour. Geology, **68**, 427–449.

Kuenen, P. H., and Perdok, W. G., 1962, Experimental abrasion 5: Frosting and defrosting of quartz grains: Jour. Geology, **70**, 648–658.

Kumar, N., and Sanders, J. E., 1974, Inlet sequences: a vertical succession of sedimentary structures and textures created by the lateral migration of tidal inlets: Sedimentology, **21**, 291–323.

Lambiase, J. J., 1980, Sediment dynamics in the macrotidal Avon River estuary, Bay of Fundy, Nova Scotia: Can. Jour. Earth Sci., **17**, 1628–1641.

Laming, D. J. C., 1966, Imbrication, paleocurrents and other sedimentary features in the lower New Red Sandstone, Devonshire, England: Jour. Sedimentary Petrology, **36**, 940–959.

Lavelle, J. W., Swift, D. J. P., Gadd, P. E., Stubblefield, W. L., Case, F. N., Brashear, H. R., and Haff, K. W., 1978, Fair weather and storm sand transport on the Long Island, New York, inner shelf: Sedimentology, **25**, 823–842.

Lavelle, J. W., Young, R. A., Swift, D. J. P., and Clarke, T. L., 1978, Near-bottom sediment concentration and fluid velocity measurements on the inner continental shelf, New York: Jour. Geophys. Res., **83**, 6052–6062.

Lawson, D. E., 1979a, Sedimentological analysis of the western

Terminus region of the Matanuska Glacier, Alaska: Hanover, NH, U.S. Army Corps of Engineers, CRREL Rept. 79-9, 112 p.

————, 1979b, A comparison of the pebble orientations in ice and deposits of the Matanuska Glacier, Alaska: Jour. Geology, **87**, 629–646.

————, 1982, Mobilization, movement and deposition of active subaerial sediment flows, Matanuska Glacier, Alaska: Jour. Geology, **90**, 279–300.

Leckie, D. A., and Walker, R. G., 1982, Storm- and tide-dominated shorelines in Cretaceous Moosebar-Lower Gates Interval-outcrop equivalents of deep basin gas trap in western Canada: Am. Assoc. Petroleum Geologists Bull., **66**, 138–157.

LeFournier, J., and Friedman, G. M., 1974, Rate of lateral migration of adjoining sea-margin sedimentary environments shown by historical records, Authie Bay, France: Geology, **2**, 497–498.

Leg 96 Scientific Staff, 1984, *Challenger* drills Mississippi Fan: Geotimes, **29**, 7, 15–18.

Leopold, L. B., Wolman, M. G., and Miller, J. P., 1964, Fluvial processes in geomorphology: San Francisco, W. H. Freeman & Co., 522 p.

Libby-French, J., 1984, Stratigraphic framework and petroleum potential of Northeastern Baltimore Canyon Trough, Mid-Atlantic Outer Continental Shelf: Am. Assoc. Petroleum Geologists Bull., **68**, 50–73.

Lindsay, J. F., Prior, D. B., and Coleman, J. M., 1984, Distributary-mouth bar development and role of submarine landslides in delta growth, South Pass, Mississippi Delta: Am. Assoc. Petroleum Geologists Bull., **68**, 1732–1743.

Link, M. H., Squires, R. L., and Colburn, I. P., 1984, Slope and deep-sea facies and paleogeography of Upper Cretaceous Chatsworth Formation, Simi Hills, California: Am. Assoc. Petroleum Geologists Bull., **68**, 850–873.

Lonsdale, P., and Malfait, B., 1974, Abyssal dunes of foraminiferal sand on the Carnegie Ridge: Geol. Soc. America Bull., **85**, 1697–1712.

Lonsdale, P., Normark, W. R., and Newman, W. A., 1972, Sedimentation and erosion on Horizon Guyot: Geol. Soc. America Bull., **82**, 289–316.

Loope, D. B., 1984, Eolian origin of Upper Paleozoic sandstones, southeastern Utah: Jour. Sedimentary Petrology, **54**, 563–580.

Lowe, D. R., 1975, Water escape structures in coarse-grained sediments: Sedimentology, **22**, 157–204.

————, 1976, Subaqueous liquefied and fluidized sediment flows and their deposits: Sedimentology, **23**, 285–308.

Lowe, D. R., and LoPiccolo, R. D., 1974, The characteristics and origins of dish and pillar structures: Jour. Sedimentary Petrology, **44**, 484–501.

Ludwick, J. C., 1970, Sand waves in the tidal entrance to Chesapeake Bay: preliminary observations: Chesapeake Sci., **11**, 98–110.

————, 1974, Tidal currents and zig-zag sand shoals in a wide estuary entrance: Geol. Soc. America Bull., **85**, 717–726.

————, 1981, Bottom sediments and depositional rates near thimble shoal channel, lower Chesapeake Bay, Virginia: Geol. Soc. America Bull., **82**, 496–506.

Lupe, R., and Ahlbrandt, T. S., 1979, Sediments of ancient eolian environments—Reservoir inhomogeneity: in McKee, E. D., ed., A study of global sand seas: U.S. Geological Survey Prof. Paper 1052, 241–252.

MacKenzie, D. B., 1972a, Primary stratigraphic traps in sandstones: in Stratigraphic oil and gas fields: Am. Assoc. Petroleum Geologists Memoir 16, 47–63.

————, 1972b, Tidal sand flat deposits in lower Cretaceous Dakota Group near Denver, Colorado: Mountain Geologist, **9**, 269–278.

MacPherson, B. A., 1978, Sedimentation and trapping mechanism in upper Miocene Stevens and older turbidite fans of southeastern San Joaquin Valley, California: Am. Assoc. Petroleum Geologists Bull., **62**, 2243–2274.

Mann, R. G., Swift, D. J. P., and Perry, R., 1981, Size classes of flow transverse bedforms in a subtidal environment, Nantucket Shoals, North American Atlantic Shelf: Geo-Marine Lettr., **1**, 39–43.

Marsaglia, K. M., and Klein, G. deV., 1983, The paleogeography of Paleozoic and Mesozoic storm depositional systems: Jour. Geology, **91**, 117–142.

Masters, C. D., 1967, Use of sedimentary structures in determination of depositional environments, Mesaverde Formation, Williams Fork Mountains, Colorado: Am. Assoc. Petroleum Geologists Bull., **51**, 2033–2043.

McBride, E. F., 1962, Flysch and associated beds of the Martinsburg Formation (Ordovician), central Appalachians: Jour. Sedimentary Petrology, **32**, 30–91.

McCave, I. N., 1970, Deposition of fine-grained suspended sediment from tidal currents: Jour. Geophys. Res., **75**, 4151–4159.

————, 1971, Wave effectiveness at the sea bed and its relationship to bedforms and deposition of mud: Jour. Sedimentary Petrology, **41**, 89–96.

————, 1979, Tidal currents at the North Hinder lightship, southern North Sea: Flow directions and turbulence in relation to maintenance of sand banks: Marine Geol., **31**, 101–114.

McDonald, D. A., and Surdam, R. C., 1984, eds., Clastic diagenesis: Am. Assoc. Petroleum Geol. Mem. 37, 434 p.

McGowen, J. H., and Groat, C. G., 1971, Van Horn Sandstone, West Texas: an alluvial fan model for mineral exploration: Texas Bureau of Econ. Geol. Rept. of Inv. **72**.

McGregor, B. A., 1977, Geophysical assessment of submarine slide northeast of Wilmington Canyon: Marine Geotech., **2**, 229–244.

McKee, E. D., 1934, The Coconino Sandstone—its history and origin: Carnegie Inst. Washington Pub. 440, 77–115.

196

————, 1945, Small-scale structures in Coconino Sandstone of northern Arizona: Jour. Geology, **53,** 313–325.

————, 1966, Structures of dunes at White Sands National Monument, New Mexico (and a comparison with structures of dunes from other selected areas): Sedimentology, **7,** 3–69.

————, 1979a, Introduction to a study of global sand seas: *in* McKee, E. D., ed., A study of global sand seas: U.S. Geol. Survey Prof. Paper 1052, 1–20.

————, 1979b, Sedimentary structures in dunes: *in* McKee, E. D., ed., A study of global sand seas: U.S. Geol. Survey Prof. Paper 1052, 83–136.

————, 1979c, Ancient sandstones considered to be eolian: *in* McKee, E. D., ed., A study of global sand seas: U.S. Geol. Survey Prof. Paper 1052, 187–240.

————, 1982, Sedimentary structures in dunes of the Namib Desert, South West Africa: Geol. Soc. America Spec. Paper 188, 64 p.

————, 1983, Eolian sand bodies of the world: *in* Brookfield, M. E., and Ahlbrandt, T. S., eds., Eolian sediments and processes: Amsterdam, Elsevier Pub. Co. Developments in Sedimentology, **38,** 1–26.

McKee, E. D., and Douglas, J. R., 1971, Growth and movement of dunes at White Sands National Monument, New Mexico: *in* Geological survey research, 1971: U.S. Geol. Survey Prof. Paper 750-D, D-108–D-114.

McKee, E. D., and Moiola, R. J., 1975, Geometry and growth of the White Sands dune field, New Mexico: U.S. Geol. Survey Jour. Res., **3,** 59–66.

McKee, E. D., Douglas, J. R., and Rittenhouse, S., 1971, Deformation of lee-side laminae in eolian dunes: Geol. Soc. America Bull., **82,** 359–378.

McLean, H., and Howell, D. G., 1984, Miocene Blanca Fan, northern Channel Islands, California: small fans reflecting tectonism and volcanism: Geo-Marine Lettr., **3,** 161–166.

Miall, A. D., 1977, A review of the braided-river depositional environment: Earth Sci. Rev., **13,** 1–62.

————, 1978a, Fluvial sedimentology: an historical review: *in* Miall, A. D., ed., Fluvial sedimentology: Can. Soc. Petroleum Geol. Mem. 5, 1–48.

————, 1978b, Lithofacies types and vertical profile models in braided river deposits: a summary: *in* Miall, A. D., ed., Fluvial sedimentology: Can. Soc. Petroleum Geol. Mem. 5, 597–604.

Middleton, G. V., 1966a, Experiments on density and turbidity currents, I: Motion of the head: Can. Jour. Earth Sci., **3,** 523–546.

————, 1966b, Experiments on density and turbidity currents, II: Uniform flow of density currents: Can. Jour. Earth Sci., **3,** 627–637.

————, 1967, Experiments on density and turbidity currents, III: Deposition of sediment: Can. Jour. Earth Sci., **4,** 475–506.

Middleton, G. V., and Hampton, M. A., 1973, Sediment gravity flows: mechanics of flow and deposition: *in* Middleton, G. V.,

and Bouma, A. H., eds., Turbidites and deep-water sedimentation: Pacific Section, Soc. Econ. Paleontologists and Mineralogists Short Course Syllabus, 1–38.

———— and ————, 1976, Subaqueous sediment transport and deposition by sediment gravity flows: *in* Stanley, D. J., and Swift, D. J. P., eds., Marine sediment transport and environmental management: New York, Wiley, 197–218.

Milliman, J. D., Pilkey, O. H., and Ross, D. A., 1972, Sediments of the continental margin off the eastern United States: Geol. Soc. America Bull., **83,** 1315–1334.

Minter, W. E. L., 1976, Detrital gold, uranium and pyrite concentrations related to sedimentology in the Precambrian Vaal Reef placer, Witwatersrand, South Africa: Econ. Geol., **71,** 157–176.

————, 1978, A sedimentological synthesis of placer gold, uranium and pyrite concentrations in Proterozoic Witwatersrand sediments: *in* Miall, A. D., ed., Fluvial sedimentology: Can. Soc. Petroleum Geol. Mem. 5, 801–830.

Moiola, R. J., and Shanmugam, G., 1984, Facies analysis of upper Jackfork Formation (Pennsylvania), DeGray Dam, Arkansas (Abs.): Am. Assoc. Petroleum Geologists Bull., **68,** 509.

Moore, D. G., 1972, Reflection profiling studies of the California Borderland: structure and Quaternary turbidite basins: Geol. Soc. America Spec. Paper 107, 142 p.

————, 1977, Submarine slides: *in* Voigt, B., ed., Rockslides and avalanches: Developments in geotechnical engineering 14a, v.1: Amsterdam, Elsevier, 563–604.

Moore, G. T., Starke, G. W., Bonham, L. C., and Woodbury, H. O., 1978, Mississippi Fan, Gulf of Mexico—physiography, stratigraphy, and sedimentation patterns: *in* Bouma, A. H., Moore, G. T., and Coleman, J. M., eds., Framework, facies, and oil-trapping characteristics of the upper continental margin: Am. Assoc. Petroleum Geologists Studies in Geology, **7,** 155–191.

Moore, P. S., and Hocking, R. M., 1983, Significance of hummocky cross-stratification in the Permian of the Carnarvon Basin, western Australia: Geol. Soc. Australia Jour., **30,** 323–331.

Morgan, J. P., ed., 1970, Deltaic sedimentation: modern and ancient: Soc. Econ. Paleontologists and Mineralogists Spec. Pub. 15, 312 p.

Morgan, J. P., Coleman, J. M., and Gagliano, S. M., 1968, Mudlumps: diapiric structures in Mississippi Delta sediments: *in* Braunstein, J., ed., Diapirs and diapirism: Am. Assoc. Petroleum Geologists Mem. 8, 145–161.

Morris, R. C., 1974a, Carboniferous rocks of the Ouachita Mountains, Arkansas: a study of facies patterns along the unstable slope and axis of a flysch trough: *in* Briggs, G., ed., Carboniferous of the southeastern United States: Geol. Soc. America Spec. Paper 148, 241–280.

————, 1974b, Sedimentary and tectonic history of the Ouachita Mountains: *in* Dickinson, W. R., ed., Tectonics and sedimentation: Soc. Econ. Paleontologists and Mineralogists Spec.

Pub. 22, 120–142.

Mossman, D. J., and Harron, G. A., 1984, Witwatersand-type paleoplacer gold in the Huronian Supergroup of Ontario, Canada: Geoscience Canada, **11,** 33–40.

Mount, J. F., 1982, Storm-surge-ebb origin of hummocky cross-stratified units of the Andrews Mountain Member, Campito Formation (lower Cambrian), White-Inyo Mountains, eastern California: Jour. Sedimentary Petrology, **52,** 941–958.

Mutti, E., 1984, The Hecho Eocene submarine fan system, south-central Pyrenees, Spain: Geo-Marine Lettr., **3,** 199–202.

Mutti, E., and Ricchi Lucchi, F., 1972, Le torbiditi dell'Appennino settentrionale: introduzione all'analisi di facies: Mem. della Soc. Geologica Italiana, **11,** 161–199.

Mutti, E., Ricchi Lucchi, F., Seguret, M., and Zanzuchi, G., 1984, Seismoturbidites: a new group of resedimented deposits: Marine Geol., **55,** 103–116.

Nami, M., 1983, Gold distribution in relation to depositional processes in the Proterozoic Carbon Leader placer, Witwatersrand, South Africa: in Collinson, J. D., and Lewin, J., eds., Modern and ancient fluvial systems: Int. Assoc. Sedimentologists Spec. Pub. 6, 563–575.

Nanz, R. H., Jr., 1954, Genesis of Oligocene sandstone reservoir, Seeligson Field, Jim Wells and Kleberg Counties, Texas: Am. Assoc. Petroleum Geologists Bull., **38,** 96–117.

Nardin, T. R., Hein, F. J., Gorsline, D. S., and Edwards, B. D., 1979, A review of mass movement processes, sediments and acoustic characteristics and contrasts in slope and base-of-slope systems versus canyon-fan-basin floor systems: in Doyle, L. E., and Pilkey, O. H., eds., Geology of continental slopes: Soc. Econ. Paleontologists and Mineralogists Spec. Pub. 27, 61–74.

Nelson, C. H., 1982, Modern shallow-water graded sand layers from storm surges, Bering Shelf: a mimic of Bouma sequences and turbidite systems: Jour. Sedimentary Petrology, **52,** 537–545.

———, 1984, The Astoria Fan: an elongate-type fan: Geo-Marine Lettr., **3,** 65–70.

Nelson, C. H., and Kulm, L. V., 1973, Submarine fans and channels: in Middleton, G. V., and Bouma, A. H., eds., Turbidites and deep-water sedimentation: Pacific Section, Soc. Econ. Paleontologists and Mineralogists Course Syllabus, 39–78.

Nelson, C. H., and Nilsen, T. H., 1974, Depositional trends of modern and ancient deep sea fans: in Dott, R. H., Jr., ed., Modern and ancient geosynclinal sedimentation: Soc. Econ. Paleontologists and Mineralogists Spec. Pub. 19, 69–91.

Nelson, C. H., Carlson, P. R., Byrne, J. V., and Alpha, T. R., 1970, Development of the Astoria Canyon-Fan physiography and comparison with similar systems: Marine Geol., **8,** 259–291.

Nelson, C. H., Dupre, W., Field, M. E., and Howard, J. E., 1982, Variation in sand body types on the eastern Bering Sea, epicontinental shelf: Geol. en Mijnb., **61,** 37–48.

Nelson, C. H., Maldonado, A., Coumes, F., Got, H., and Mon-

aco, A., 1984, The Ebro deep-sea fan system: Geo-Marine Lettr., **3,** 125–133.

Niino, H., and Emery, K. O., 1961, Sediments of shallow portions of East China Sea and South China Sea: Geol. Soc. America Bull., **72,** 731–762.

Nijman, W., and Puigdefabregas, C., 1978, Coarse-grained point bar structure in a molasse-type fluvial system, Eocene Castisent Sandstone Formation, South Pyrenean Basin: in Miall, A. D., ed., Fluvial sedimentology: Can Soc. Petroleum Geol. Mem. 5, 487–510.

Nilsen, T. H., 1969, Old Red sedimentation in Buelandet-Vaerlandet Devonian district, western Norway: Sediment. Geol., **3,** 35–57.

———, 1984a, Submarine fan facies associations of the Eocene Butano Sandstone, Santa Cruz Mountains: Geo-Marine Lettr., **3,** 167–172.

———, 1984b, Trench-fill submarine-fan facies associations of the upper Cretaceous Chugach terraine, southern Alaska: Geo-Marine Lettr., **3,** 179–186.

Nilsen, T. H., and Abbate, E., 1984, Submarine-fan facies associations of the upper Cretaceous and Paleocene Gottera Sandstone, Liguran Apennines, Italy: Geo-Marine Lettr., **3,** 193–198.

Nio, S. D., 1976, Marine transgressions as a factor in formation of sandwave complexes: Geol. en Mijnb., **55,** 18–40.

Nio, S. D., and Nelson, C. H., 1982, The North Sea and northeastern Bering Sea: a comparative study of the occurrence and geometry of sand bodies of two shallow epicontinental shelves: Geol. en Mijnb., **61,** 105–114.

Nio, S. D., Siegenthaler, C., and Yang, C. S., 1983, Megaripple cross-bedding as a tool for the reconstruction of the paleohydraulics in a Holocene subtidal environment, SW Netherlands: Geol. en Mijnb., **62,** 499–510.

Normark, W. R., 1969, Growth patterns of deep-sea fans: Am. Assoc. Petroleum Geologists Bull., **54,** 2170–2195.

———, 1978, Fan valleys, channels and depositional lobes on modern submarine fans: characteristics for recognition of sandy turbidite environments: Am. Assoc. Petroleum Geologists Bull., **62,** 912–931.

Normark, W. R., and Barnes, N. E., 1984, Aftermath of COMFAN: comments, not solutions: Geo-Marine Lettr., **3,** 223–224.

Normark, W. R., and Gutmacher, C. E., 1984, Delgada Fan: preliminary interpretation of channel development: Geo-Marine Lettr., **3,** 79–84.

Normark, W. R., and Piper, D. J. W., 1972, Sediments and growth pattern of Navy deep-sea fan, San Clemente Basin, California borderland: Jour. Geology, **80,** 198–223.

——— and ———, 1984, Navy Fan, California borderland: growth pattern and depositional processes: Geo-Marine Lettr., **3,** 101–108.

Normark, W. R., Barnes, N. E., and Coumes, F., 1984, Rhone deep sea fan: a review: Geo-Marine Lettr., **3,** 155–160.

Normark, W. R., Mutti, E., and Bouma, A. H., 1984, Problems

in turbidite research: a need for COMFAN: Geo-Marine Lettr., **3**, 53–56.

Normark, W. R., Piper, D. J. W., and Hess, G. R., 1979, Distributary channels, sand lobes, and mesotopography of Navy submarine fan, California borderland, with applications to ancient fan sediments: Sedimentology, **26**, 749–774.

Normark, W. R., Piper, D. J. W., and Stow, D. A. V., 1983, Quaternary development of channels, levees, and lobes on middle Laurentian Fan: Am. Assoc. Petroleum Geologists Bull., **67**, 1400–1409.

Normark, W. R., Gutmacher, C. E., Chase, T. E., and Wilde, P., 1984, Monterey Fan: growth pattern control by basin morphology and changing sea levels: Geo-Marine Lettr., **3**, 93–100.

Oertel, G. F., 1972, Sediment transport of estuary entrance shoals and the formation of swash platforms: Jour. Sedimentary Petrology, **42**, 858–864.

Oertel, G. F., and Howard, J. D., 1972, Water circulation and sedimentation at estuary entrances on the Georgia Coast: *in* Swift, D. J. P., Duane, D. B., and Pilkey, O. H., eds., Shelf sediment transport: Stroudsburg, Pennsylvania, Dowden, Hutchinson, and Ross, 411–428.

Off, T., 1963, Rhythmic linear sand bodies caused by tidal currents: Am. Assoc. Petroleum Geologists Bull., **47**, 324–341.

Ojakangas, R. W., 1982, Tidal deposits in the early Proterozoic basin of the Lake Superior region—The Palms and Pkegama formations: evidence for subtidal shelf deposition of Superior-type banded iron-formation: *in* Medeiras, G., ed., Geol. Soc. America Mem. 160, 49–66.

Oomkens, E., 1974, Lithofacies relations in the Late Quaternary Niger Delta complex: Sedimentology, **21**, 195–221.

Otvos, E. G., 1970, Development and migration of barrier islands, northern Gulf of Mexico: Geol. Soc. America Bull., **81**, 341–348.

Ovenshine, A. T., Lawson, D. E., and Bartsch-Winkler, S. R., 1976, The Placer River Silt—an intertidal deposit caused by the 1964 Alaska Earthquake: U.S. Geol. Survey Jour. Res., **4**, 141–162.

Ovenshine, A. T., Bartsch-Winkler, S. R., O'Brien, N. R., and Lawson, D. E., 1975, Sediments of the high tidal range environment of Upper Turnagain Arm, Alaska: *in* Recent and ancient sedimentary environments in Alaska: Alaska Geol. Soc., 1–40.

Padgett, G., and Ehrlich, R., 1978, An analysis of two tectonically controlled drainage nets of mid-Carboniferous age in southern West Virginia: *in* Miall, A. D., ed., Fluvial sedimentology: Can. Soc. Petroleum Geol. Mem. 5, 789–800.

Paine, W. R., 1970, Petrology and sedimentation of the Hackberry sequences of southwest Louisiana: Gulf Coast Assoc. Geol. Soc. Trans., **20**, 37–55.

Palomino-Cardenas, J. R., 1976, Sedimentological and environmental study of the Fluvio-deltaic Cabo Blanco Sandstone Member, Echinocyamus Formation, Lower Eocene, Talara

Basin, N.W. Peru: Unpub. Ph.D. dissertation, Univ. of Illinois at Urbana-Champaign, 113 p.

Park, Y. A., and Song, M. Y., 1971, Sediments of the continental shelf off the southern coasts of Korea: Ocean. Soc. Korea Jour., **6**, 16–24.

Payne, M. Y., 1976, Basinal sandstone facies, Delaware Basin, west Texas and southwest New Mexico: Am. Assoc. Petroleum Geologists Bull., **60**, 515–527.

Pestrong, R., 1972, Tidal flat sedimentation at Cooley Landing, southwest San Francisco Bay: Sediment. Geol., **8**, 251–288.

Piper, D. J. W., 1970, Transport and deposition of Holocene sediment on LaJolla deep sea fan, California: Marine Geol., **8**, 211–227.

Piper, D. J. W., and Normark, W. R., 1983, Turbidite depositional patterns and flow characteristics, Navy submarine fan, California borderland: Sedimentology, **30**, 681–694.

Piper, D. J. W., Stow, D. A. V., and Normark, W. R., 1984, The Laurentian Fan: Sohm Abyssal Plain: Geo-Marine Lettr., **3**, 141–146.

Plint, A. G., 1983, Sandy fluvial point-bar sediments from the Middle Eocene of Dorset, England: *in* Collinson, J. D., and Lewin, J., eds., Modern and ancient fluvial systems: Int. Assoc. Sedimentologists Spec. Pub. 6, 355–368.

Pollard, J. E., Steel, R. J., and Undersrud, E., 1982, Facies sequences and trace fossils in lacustrine-fan delta deposits, Hornelen Basin (M. Devonian), western Norway: Sediment. Geol., **32**, 63–87.

Postma, G., 1983, Water escape structures in the context of a depositional model of a mass flow-dominated conglomeratic fan-delta (Abrioja Formation, Pliocene, Almeria Basin, SE Spain): Sedimentology, **30**, 19–103.

———, 1984, Slumps and their deposits in fan delta front and slope: Geology, **12**, 27–30.

Potter, P. E., and Pryor, W. A., 1961, Dispersal centers of Paleozoic and later clastics of the upper Mississippi Valley and adjacent areas: Geol. Soc. America Bull., **72**, 1195–1250.

Pretorius, D. A., 1974, Gold in the Proterozoic sediments of South Africa: systems, paradigms and models: Univ. of Witwatersrand Econ. Geol. Research Unit, Info. Cir. 87, 22 p.

Prior, D. B., and Coleman, J. M., 1977, Disintegrating retrogressive landslides on very low-angle subaqueous slopes, Mississippi Delta: Marine Geotech., **3**, 37–60.

——— and ———, 1978, Submarine landslides on the Mississippi River delta-front slope: Geology and Man, **19**, 41–53.

Prior, D. B., and Suhayda, J. N., 1979, Application of infinite stress slope analysis to subaqueous sediment instability, Mississippi Delta: Engin. Geol., **14**, 1–10.

Pryor, W. A., 1961, Sand trends and paleoslope in Illinois Basin and Mississippi Embayment: *in* Peterson, J. A., and Osmond, J. C., eds., Geometry of sandstone bodies: Am. Assoc. Petroleum Geologists, 119–133.

Puigdefabregas, C., and Van Vliet, A., 1978, Meandering stream deposits from the Tertiary of the southern Pyrenees: *in* Miall,

A. D., ed., Fluvial sedimentology: Can. Soc. Petroleum Geol. Mem. 5, 469–487.

Putnam, P. E., 1982, Fluvial channel sandstones within upper Mannville (Albian) of the Lloydminster area, Canada: geometry, petrography, and paleographic implications: Am. Assoc. Petroleum Geologists Bull., **66**, 436–456.

Putnam, P. E., and Oliver, T. A., 1980, Stratigraphic traps in channel sandstones in the upper Mannville (Albian) of east-central Alberta: Bull. Can. Petrol. Geol., **28**, 489–508.

Rackley, R. I., 1972, Environment of Wyoming Tertiary uranium deposits: Am. Assoc. Petroleum Geologists Bull., **56**, 755–774.

———, 1975, Problems of converting potential uranium resources into minable reserves: Mineral resources and the environment: Washington, DC, National Academy of Sciences, 120–140.

Reading, H. G., ed., 1978, Sedimentary environments and facies: Oxford, Blackwell Pub. Co., 557 p.

Redfield, A. C., 1958, The influence of the continental shelf on the tides of the Atlantic coast of the United States: Jour. Marine Res., **17**, 432–448.

Reiche, P., 1938, An analysis of cross-lamination—the Coconino Sandstone: Jour. Geology, **46**, 905–932.

Reineck, H. E., 1955, Haftrippeln and haftwarzen, ablagerungsformen von flugsand: Senck. Lethaea, **36**, 347–357.

———, 1963, Sedimentgefuge im Bereich der Sudliche Nordsee: Abhandle. Sencken. Nat. Gesell., **505**, 1–138.

———, 1967, Layered sediments of tidal flats, beaches, and shelf bottoms of the North Sea: in Lauff, G. H., ed., Estuaries: Am. Assoc. Adv. Sci. Spec. Pub. 83, 191–200.

———, 1972, Tidal flats: in Rigby, J. K., and Hamblin, W. K., eds., Recognition of ancient sedimentary environments: Soc. Econ. Paleontologists and Mineralogists Spec. Pub. 16, 146–159.

Reinert, S. L., and Davies, D. K., 1975, Third Creek Field, Colorado: a study of sandstone environments and diagenesis: Mountain Geologist, **13**, 47–60.

Ricchi Lucchi, F., Colella, A., Gabbianella, G., Rossi, S., and Normark, W. R., 1984, The Crati submarine fan, Ionian Sea: Geo-Marine Lettr., **3**, 71–78.

Roberts, H. H., Adams, R. D., and Cunningham, R. H. W., 1980, Evolution of sand-dominant subaerial phase, Atchafalaya Delta, south-central Louisiana: Am. Assoc. Petroleum Geologists Bull., **64**, 264–279.

Rust, B. R., 1972, Structures and processes in a braided river: Sedimentology, **18**, 1171–1178.

———, 1978, Depositional models for braided alluvium: in Miall, A. D., ed., Fluvial sedimentology: Can. Soc. Petroleum Geol. Mem. 5, 605–626.

———, 1981, Sedimentation in an arid-zone anastomosing fluvial system: Cooper's Creek, central Australia: Jour. Sedimentary Petrology, **51**, 745–756.

Rust, B. R., and Legun, A. S., 1983, Modern anastomosing-fluvial deposits in arid central Australia, and a Carboniferous

analogue in New Brunswick, Canada: in Collinson, J. D., and Lewin, J., eds., Modern and ancient fluvial systems: Int. Assoc. Sed. Spec. Pub., **6**, 385–392.

Ryer, T. A., 1977, Patterns of Cretaceous shallow-marine sedimentation, Coalville and Rockport areas, Utah: Geol. Soc. America Bull., **88**, 177–188.

———, 1983, Transgressive-regressive cycles and the occurrence of coal in some Upper Cretaceous strata of Utah: Geology, **11**, 207–210.

Sabins, F. F., Jr., 1962, How do Bisti and Dead Horse Creek strata traps compare?: Oil and Gas Jour., August 13, 1962, 6 p.

Sakamoto-Arnold, C. M., 1981, Eolian features produced by the December 1977 windstorm, southern San Joaquin Valley, California: Jour. Geology, **89**, 129–137.

Sangree, J. B., and Widmier, J. M., 1977, Seismic stratigraphy and global changes of sea level, Part 9: Seismic interpretation of clastic depositional facies: in Payton, C. E., ed., Seismic stratigraphy: Am. Assoc. Petroleum Geologists Mem. 26, 165–184.

Saxena, R. S., 1976a, Modern Mississippi Delta—depositional environments and processes: Am. Assoc. Petroleum Geologists Guidebook, New Orleans Mtg., 125 p.

———, 1976b, Deltas: in Saxena, R. S., ed., Sedimentary environments and hydrocarbons: Am. Assoc. Petroleum Geologists Short Course Syllabus, New Orleans Mtg., 217 p.

———, 1976c, Sand bodies and sedimentary environments of the modern Mississippi Delta—an excellent model for exploration in deltaic sandstone reservoirs: Cairo, Egyptian Petroleum Corp. Exploration Seminar Syllabus, 53 p.

Schlager, W., and Chermak, A., 1979, Sediment facies of plateau-basin transition, Tongue of the Ocean, Bahamas: in Doyle, L. J., and Pilkey, O. H., Jr., eds., Geology of continental slopes: Soc. Econ. Paleontologists and Mineralogists Spec. Pub. 27, 193–208.

Schlee, J. S., 1981, Seismic stratigraphy of Baltimore Canyon Trough: Am. Assoc. Petroleum Geologists Bull., **65**, 26–53.

Schlee, J. S., and Moench, R. H., 1961, Properties and genesis of "Jackpile" Sandstone, Laguna, New Mexico: in Peterson, J. A., and Osmond, J. C., eds., Geometry of sandstone bodies: Tulsa, Am. Assoc. Petroleum Geologists, 134–150.

Schlee, J. S., Dillon, W. P., and Grow, J. A., 1979, Structure of the continental slope off the eastern United States: in Doyle, L. E., and Pilkey, O. H., eds., Geology of continental slopes: Soc. Econ. Paleontologists and Mineralogists Spec. Pub. 27, 95–118.

Scholle, P. A., and Schluger, P. R., eds., 1979, Aspects of diagenesis: Soc. Econ. Paleontologists and Mineralogists Spec. Pub. 26, 443 p.

Schramm, M. W., Jr., Dedman, F. V., and Lindsey, P., 1977, Practical stratigraphic modelling and interpretation: in Payton, C. E., ed., Seismic stratigraphy: Am. Assoc. Petroleum Geologists Mem. 26, 477–502.

Schumm, S. A., 1968, Speculations concerning paleohydrologic controls of terrestrial sedimentation: Geol. Soc. America Bull., **79,** 1573–1588.

Schwartz, M. L., 1971, The multiple causality of barrier islands: Jour. Geology, **79,** 91–94.

Scruton, P. C., 1956, Oceanography of Mississippi Delta sedimentary environments: Am. Assoc. Petroleum Geologists Bull., **40,** 2864–2952.

————, 1960, Delta building and the deltaic sequence: *in* Shepard, F. P., Phleger, F. B., and Van Andel, T. H., eds., Recent sediments, northwest Gulf of Mexico: Am. Assoc. Petroleum Geologists, 82–102.

Seeber, L., and Gornitz, V., 1983, River profiles along the Himalayan Arc as indicators of active tectonics: Tectonophysics, **92,** 335–367.

Selley, R. C., 1976, Subsurface environmental analysis of North Sea sediments: Am. Assoc. Petroleum Geologists Bull., **60,** 184–195.

Sellwood, B. W., 1972, Tidal-flat sedimentation in the lower Jurassic of Bornholm, Denmark: Paleogeog., Paleoclim., Paleoecol., **11,** 93–106.

Semeniuk, V., 1981, Sedimentology and stratigraphic sequence of a tropical tidal flat, northwestern Australia: Sediment. Geol., **29,** 195–221.

Seyler, B., 1984, Role of diagenesis in formation of stratigraphic traps in Aux Cases of Illinois Basin (Abs.): Am. Assoc. Petroleum Geologists Bull., **68,** 526–527.

Shanmugam, G., and Moiola, R. J., 1982, Eustatic control of turbidites and winnowed turbidites: Geology, **10,** 231–235.

Shannon, J. P., and Dahl, A. R., 1971, Deltaic stratigraphic traps in West Tuscola Field, Taylor County, Texas: Am. Assoc. Petroleum Geologists Bull., **55,** 1194–1205.

Sharp, R. P., 1963, Wind ripples: Jour. Geology, **71,** 617–636.

————, 1966, Kelso dunes, Mohave Desert, California: Geol. Soc. America Bull., **77,** 1045–1073.

————, 1964, Wind-driven sand in Coachella Valley, California: Geol. Soc. America Bull., **75,** 785–804.

————, 1979, Intradune flats of the Algodones Chain, Imperial Valley, California: Geol. Soc. America Bull., **90,** 908–916.

Shepard, F. P., 1932, Sediments of the continental shelves: Geol. Soc. America Bull., **43,** 1017–1039.

————, 1960, Rise of sea level along northwest Gulf of Mexico: *in* Shepard, F. P., Phleger, F. B., and Van Andel, Tj. H., 1960, Recent sediments northwest Gulf of Mexico: Tulsa, Am. Assoc. Petroleum Geol., 338–344.

————, 1963, Submarine geology, 2d ed.: New York, Harpers, 557 p.

————, 1973, Sea floor of Magdalena Delta and Santa Maria area, Colombia: Geol. Soc. America Bull., **84,** 1955–1972.

————, 1976, Tidal components of currents in submarine canyons: Jour. Geology, **84,** 343–350.

Shepard, F. P., and Marshall, N. F., 1973, Currents along floors of submarine canyons: Am. Assoc. Petroleum Geologists Bull., **57,** 244–264.

Shepard, F. P., Dill, R. F., and Von Rad, U., 1969, Physiography and sedimentary processes of LaJolla submarine fan and fan-valley, California: Am. Assoc. Petroleum Geologists Bull., **53,** 390–420.

Shepard, F. P., Marshall, N. F., McLoughlin, P. A., and Sullivan, G. G., 1979, Currents in submarine canyons and other sea valleys: Am. Assoc. Petroleum Geologists Studies in Geology, **8,** 173 p.

Sherborne, J. E., Jr., Buckovic, W. A., DeWitt, D. B., Hellinger, T. S., and Pavlak, S. J., 1979, Major uranium discovery in volcaniclastic sediments, Basin and Range Province, Yavapai County, Arizona: Am. Assoc. Petroleum Geologists Bull., **63,** 621–646.

Sheridan, R. E., Golovchenko, X., and Ewing, J. I., 1974, Late Miocene turbidite horizon in Blake-Bahama Basin: Am. Assoc. Petroleum Geologists Bull., **58,** 1797–1805.

Shirley, M. L., 1966, Deltas: Houston, Houston. Geol. Soc., 251 p.

Shotton, F. W., 1937, The Lower Bunter sandstones of north Worchestershire and east Shropshire (England): Geol. Mag., **74,** 534–553.

Smith, D. G., 1983, Anastomosed fluvial deposits: modern examples from western Canada: *in* Collinson, J. D., and Lewin, J., eds., Modern and ancient fluvial systems: Int. Assoc. Sedimentologists Spec. Pub. 6, 155–168.

Smith, D. G., and Putnam, P. E., 1980, Anastomosed river deposits: modern and ancient examples in Alberta, Canada: Can. Jour. Earth Sci., **17,** 1396–1406.

Smith, D. G., and Smith, N. D., 1980, Sedimentation in anastomozed river systems: examples from alluvial valleys near Banff, Alberta: Jour. Sedimentary Petrology, **50,** 157–164.

Smith, J. D., 1969, Geomorphology of a sand ridge: Jour. Geology, **77,** 39–55.

Smith, N. D., 1970, The braided stream depositional environment: comparison of the Platte River with some Silurian clastic rocks, north-central Appalachians: Geol. Soc. America Bull., **81,** 2993–3014.

————, 1971, Transverse bars and braiding in the Lower Platte River, Nebraska: Geol. Soc. America Bull., **82,** 3407–3420.

Smith, N. D., and Minter, W. E. L., 1980, Sedimentological controls of gold and uranium in two Witwatersrand paleoplacers: Econ. Geol., **75,** 1–15.

Smith, N. D., and Smith, D. G., 1984, William River: an outstanding example of channel widening and braiding caused by bed-load addition: Geology, **12,** 78–82.

Song, W., Yoo, D., and Dyer, K. R., 1983, Sediment distribution, circulation and provenance in a macrotidal bay: Garolin Bay, Korea: Marine Geol., **52,** 121–140.

Sonnenberg, S. A., 1977, Interpretation of Cotton Valley depositional environment from core study, Frierson Field, Louisiana: Gulf Coast Assoc. Geol. Soc. Trans., **27,** 320–325.

Spearing, D. R., 1975, Summary sheets of sedimentary deposits: Geol. Soc. America Pub. MC-8.

Stanley, K. O., and Surdam, R. C., 1978, Sedimentation on the front of Eocene Gilbert-type deltas, Washakie Basin, Wyoming: Jour. Sedimentary Petrology, **48**, 557–573.

Stauffer, P. H., 1967, Grain-flow deposits and their implications. Santa Ynez Mountains, California: Jour. Sedimentary Petrology, **37**, 487–508.

Steel, R. J., 1976, Devonian basins of western Norway—Sedimentary response to tectonism and to varying tectonic context: Tectonophysics, **36**, 207–224.

Steel, R. J., Maehle, S., Nilsen, H., Roe, S. L., and Spinnangr, A., 1977, Coarsening-upward cycles in the alluvium of Hornelen Basin (Devonian), Norway: sedimentary response to tectonic events: Geol. Soc. America Bull., **88**, 1124–1134.

Steidtmann, J. R., 1973, Ice and snow in eolian sand dunes of southwestern Wyoming: Science, **179**, 796–798.

———, 1974, Evidence for eolian cross-stratification in sandstone of the Casper Formation, southernmost Laramie Basin, Wyoming: Geol. Soc. America Bull., **85**, 1835–1842.

Stow, D. A. V., 1981, Laurentian Fan: morphology, sediments, processes and growth pattern: Am. Assoc. Petroleum Geologists Bull., **65**, 375–393.

Stow, D. A. V., Howell, D. G., and Nelson, C. H., 1984, Sedimentary, tectonic, and sea level controls on submarine fan and slope-apron turbidite systems: Geo-Marine Lettr., **3**, 57–64.

Stride, A. H., 1963, Current-swept sea floors near the southern half of Great Britain: Geol. Soc. London Quar. Jour., **119**, 175–199.

———, ed., 1982, Offshore tidal sands: London, Chapman and Hall, 222 p.

Stuart, C. J., and Caughey, C. A., 1977, Seismic facies and sedimentology of terrigeneous Pleistocene deposits in northwest and central Gulf of Mexico: in Payton, C. E., ed., Seismic stratigraphy: Am. Assoc. Petroleum Geologists Mem. 26, 249–276.

Stubblefield, W. L., Kersey, D. G., and McGrail, D. W., 1983, Development of middle continental shelf sand ridges: New Jersey: Am. Assoc. Petroleum Geologists Bull., **67**, 817–830.

Sundborg, A., 1956, The River Klaralven: Geog. Annaler, **38**, 127–316.

Sverdrup, H. U., Johnson, M. W., and Fleming, R. H., 1942, The oceans: Englewood Cliffs, New Jersey, Prentice-Hall, 1087 p.

Swift, D. J. P., 1968, Coastal erosion and transgressive stratigraphy: Jour. Geology, **76**, 444–456.

———, 1970, Quaternary shelves and the return to grade: Marine Geol., **8**, 5–30.

———, 1974, Continental shelf sedimentation: in Burk, C. A., and Drake, C. L., 1974, The geology of continental margins: New York, Springer-Verlag, 117–136.

Swift, D. J. P., and Field, M. E., 1981, Evolution of a classic sand ridge field: Maryland sector, North American inner shelf: Sedimentology, **28**, 461–482.

Swift, D. J. P., and Freeland, G. L., 1978, Current lineation and sand waves on the inner shelf, Middle Atlantic Bight of North America: Jour. Sedimentary Petrology, **48**, 1257–1266.

Swift, D. J. P., Freeland, G. L., and Young, R. A., 1979, Time and space distribution of megaripples and associated bedforms, Middle Atlantic Bight, North American Atlantic Shelf: Sedimentology, **26**, 389–406.

Swift, D. J. P., Stanley, D. J., and Curray, J. R., 1971, Relict sediments on continental shelves: a reconsideration: Jour. Geology, **79**, 322–346.

Swift, D. J. P., Figueiredo, A. G., Jr., Freeland, G. L., and Oertel, G. F., 1983, Hummocky cross-stratification and megaripples: a geological double standard: Jour. Sedimentary Petrology, **53**, 1295–1317.

Swift, D. J. P., Nelson, T., McHone, J. H., Holliday, B., Palmer, H., and Shideler, G., 1977, Holocene evolution of the inner shelf of southern Virginia: Jour. Sedimentary Petrology, **47**, 1454–1474.

Swift, D. J. P., Parker, G., Lanfredi, N. W., Perillo, G., and Figge, K., 1978, Shoreface-connected sand ridges on American and European shelves: a comparison: Estuaries and Coastal Mar. Sci., **7**, 257–273.

Swift, D. J. P., Sears, P. C., Bohlke, B., and Hunt, R., 1978, Evolution of a shoal retreat massif, North Carolina shelf: inferences from areal geology: Marine Geol., **27**, 19–24.

Swift, D. J. P., Young, R. A., Clarke, T. L., Vincent, C. E., Niederoda, A., and Lesht, B., 1981, Sediment transport in the Middle Atlantic Bight of North America: synopsis of recent observations: in Nio, S. D., Schuttenhelm, R. T. E., and Van Weering, T. C. E., eds., Holocene marine sedimentation in the North Sea Basin: Int. Assoc. Sedimentologists Spec. Pub. 5, 361–384.

Swinbanks, D. D., 1982, Intertidal exposure zones: a way to subdivide the shore: Jour. Exp. Mar. Biol. Ecol., **62**, 69–86.

Swinbanks, D. D., and Murray, J. W., 1981, Biosedimentological zonation of Boundary Bay tidal flats, Fraser River Delta, British Columbia: Sedimentology, **28**, 201–237.

Taira, A., and Scholle, P. A., 1979, Deposition of resedimented sandstone beds in Pico Formation, Ventura Basin, California, as interpreted from magnetic fabric measurements: Geol. Soc. America Bull., **90**, 952–962.

Tankard, A. J., and Hobday, D. K., 1977, Tide-dominated back barrier sedimentation, early Ordovician Cape Basin, Cape Peninsula, South Africa: Sediment. Geol., **18**, 135–160.

Taylor, J. C. M., 1977, Sandstones as reservoir rocks: in Hobson, G. S., ed., Developments in petroleum geology, I: London, Applied Science Pub., Ltd., 147–196.

Ten Haaf, E., 1959, Graded beds of the northern Appenines: Ph.D. dissertation, Univ. of Groningen, 102 p.

Terwindt, J. H. J., 1971, Sand waves in the southern bight of the North Sea: Marine Geol., **10**, 51–68.

Teyssen, T. A. L., 1984, Sedimentology of the Minette-oolitic ironstones of Luxembourg and Lorraine: a Jurassic subtidal sandwave complex: Sedimentology, **31**, 195–211.

Thompson, D. B., 1969, Dome-shaped aeolian dunes in the Frodsham Member of the so-called "Keuper" Sandstone Formation (Scythian-Anisian; Triassic) at Frodsham, Cheshire (England): Sediment. Geol., **3**, 263–289.

Thompson, R. W., 1968, Tidal flat sedimentation on the Colorado River Delta, northwest Gulf of California: Geol. Soc. America Mem. 107, 133 p.

Tillman, R. W., and Martinsen, R. S., 1984, The Shannon shelf-ridge sandstone complex, Salt Creek Anticline area, Powder River Basin, Wyoming: *in* Tillman, R. D., and Siemers, C. T., eds., Siliciclastic shelf sediments: Soc. Econ. Paleontologists and Mineralogists Spec. Pub. 34, 85–142.

Tillman, R. W., Hughes, S. M., Scott, R. M., and Dailey, D. H., 1981, Upper Cretaceous deep-water Winters Sandstone, Cities Service Nixon Community No. 1, Solano County, California: *in* Siemers, C. T., Tillman, R. W., and Williamson, C. R., eds., Deep water clastic sediments: a core workshop: Soc. Econ. Paleontologists and Mineralogists Core Workshop No. 2, 47–76.

Tsaor, H., 1983, Dynamics processes acting on longitudinal (sief) sand dune: Sedimentology, **30**, 567–578.

Tucholke, B. E., Vogt, P. R., et al., 1979, Initial reports of the Deep Sea Drilling Project, v. 44: Washington, DC, U.S. Government Printing Office, 115 p.

Turner, J. R., and Conger, S. J., 1984, Environment of deposition and reservoir properties of the Woodbine Sandstone at Kurten Field, Brazos County, Texas: *in* Tillman, R. W., and Siemers, C. T., eds., Siliciclastic shelf sediments: Soc. Econ. Paleontologists and Mineralogists Spec. Pub. 34, 215–250.

Tyler, N., and Ethridge, F. G., 1983, Fluvial architecture of Jurassic uranium-bearing sandstones, Colorado Plateau, western United States: *in* Collinson, J. D., and Lewin, J., eds., Modern and ancient fluvial systems: Int. Assoc. Sedimentologists Spec. Pub. 6, 533–548.

Vail, P. R., and Hardenbol, J., 1979, Sea-level changes during the Tertiary: Oceanus, **22**, 71–79.

Vail, P. R., Mitchum, R. M., Jr., and Thompson, S., III, 1977, Seismic stratigraphy and global changes in sea level, Part 4: Global cycles of relative changes of sea level: *in* Payton, C. E., ed., Seismic stratigraphy—applications to hydrocarbon exploration: Am. Assoc. Petroleum Geol. Mem. 26, 83–97.

Van Der Meulen, S., 1982, The sedimentary facies and setting of Eocene point bar deposits, Monllobat Formation, southern Pyrenees, Spain: Geol. en Mijnb., **16**, 217–227.

Van Straaten, L. M. J. U., 1952, Biogene textures and the formation of shell beds in the Dutch Wadden Sea: Koninkl. Nederlandse Akad. Wetensche. Proc., **55**, Ser. B., 500–516.

———, 1953a, Rhythmic pattern on Dutch North Sea beaches: Geol. en Mijnb., **15**, 31–43.

———, 1953b, Megaripples in the Dutch Wadden Sea and in the Basin of Arachon (France): Geol. en Mijnb., **15**, 1–11.

———, 1954, Sedimentology of Recent tidal flat deposits and the Psammites du Condroz: Geol. en Mijnb., **16**, 25–47.

———, 1959, Minor structures of some Recent littoral and neritic sediments: Geol. en Mijnb., **21**, 197–216.

———, 1961, Sedimentation in tidal flat areas: Alberta Soc. Petroleum Geol. Jour., **9**, 203–226.

———, 1965, Coastal barrier deposits in south and north Holland: Meded. van Geologische Stichting, **17**, 41–75.

Van Veen, J., 1935, Sandwaves in the southern North Sea: Hydrograph. Rev., **12**, 21–29.

Vincelette, R. R., and Chittum, W. F., 1981, Exploration of oil accumulation in Entrada Sandstone, San Juan Basin, New Mexico: Am. Assoc. Petroleum Geologists Bull., **65**, 2546–2570.

Visher, G. S., 1965, Use of the vertical profile in environmental reconstruction: Am. Assoc. Petroleum Geologists Bull., **49**, 41–61.

———, 1972, Physical characteristics of fluvial deposits: *in* Rigby, J. K., and Hamblin, W. K., eds., Recognition of ancient sedimentary environments: Soc. Econ. Paleontologists and Mineralogists Spec. Pub. 16, 84–97.

Visher, G. S., Saitta, D., and Phillips, R. S., 1971, Pennsylvanian delta patterns and petroleum occurrences in eastern Oklahoma: Am. Assoc. Petroleum Geologists Bull., **55**, 1206–1230.

Visser, M. J., 1980, Neap-spring cycles reflected in Holocene subtidal large-scale bedform deposits: a preliminary note: Geology, **8**, 543–546.

Von Bruun, V., and Hobday, D. K., 1976, Early Precambrian tidal sedimentation in the Pongola Supergroup of South Africa: Jour. Sedimentary Petrology, **46**, 670–679.

Von Der Borch, C. C., Sclater, J. G., et al., 1974, Site 218: Initial report of the Deep Sea Drilling Project, v. 22: Washington, DC: U.S. Government Printing Office, 325–332.

Vos, R. G., 1975, An alluvial plain and lacustrine model for the Precambrian Witwatersrand deposits of South Africa: Jour. Sedimentary Petrology, **45**, 480–493.

Wageman, J. M., Hilde, T. W. C., and Emery, K. O., 1970, Structural framework of East China Sea and Yellow Sea: Am. Assoc. Petroleum Geologists Bull., **65**, 1611–1643.

Walker, R. G., 1966, Shale Grit and Grindslow Shales: transition from turbidite to shallow-water sediments in the Upper Carboniferous of northern England: Jour. Sedimentary Petrology, **36**, 9–114.

———, 1967, Turbidite sedimentary structures and their relationship to proximal and distal depositional environments: Jour. Sedimentary Petrology, **37**, 25–43.

———, 1973, Mopping up the turbidite mess. A history of the turbidity current concept: *in* Ginsburg, R. N., ed., Evolving concepts in sedimentology: Baltimore, Johns Hopkins Univ. Press, 1–37.

———, 1975, Nested submarine-fan channels in the Capistrano Formation, San Clemente, California: Geol. Soc. America Bull., **86**, 915–924.

———, 1976, Facies Models 1: General introduction: Geoscience Canada, **3**, 21–24.

———, 1978, Deep-water sandstone facies and ancient submarine fans: models for exploration for stratigraphic traps: Am. Assoc. Petroleum Geologists Bull., **62**, 932–966.

———, 1983a, Cardium Formation 1. "Cardium, a turbidity current deposit" (Beach, 1955): a brief history of ideas: Bull. Can. Petrol. Geol., **31**, 205–212.

———, 1983b, Cardium Formation 2. Sand-body geometry and stratigraphy in the Garrington-Caroline-Ricinus area, Alberta—The "ragged blanket" model: Bull. Can. Petrol. Geol., **31**, 14–26.

———, 1983c, Cardium Formation 3. Sedimentology and stratigraphy in the Garrington-Caroline area, Alberta: Bull. Can. Petrol. Geol., **31**, 213–230.

Walker, T. R., 1967, Formation of red beds in modern and ancient deserts: Geol. Soc. America Bull., **78**, 353–368.

———, 1975, Intrastratal sources of uranium in first-cycle, nonmarine red beds (Abs.): Am. Assoc. Petroleum Geologists Bull., **59**, 925.

Walker, T. R., and Harms, J. C., 1972, Eolian origin of flagstone beds, Lyons Sandstone (Permian), type area, Boulder County, Colorado: Mountain Geologist, **9**, 279–288.

Wanless, H. R., 1981, Environments and dynamics of clastic sediment dispersal across Cambrian of Grand Canyon (Abs.): Am. Assoc. Petroleum Geol. Bull., **65**, 1005.

Wanless, H. R., et al., 1970, Late Paleozoic deltas in the central and eastern United States: in Morgan, J. P., ed., Deltaic sedimentation: modern and ancient: Soc. Econ. Paleontologists and Mineralogists Spec. Pub. 15, 215–245.

Wasson, R. J., and Hyde, R., 1983, Factors determining desert dune type: Nature, **304**, 337–339.

Weaver, P. P. E., and Kuijpers, A., 1983, Climatic control of turbidite deposition on the Madeira Abyssal Plain: Nature, **306**, 360–363.

Webb, G. W., 1981, Stevens and earlier Miocene turbidite sandstones, southern San Joaquin Valley, California: Am. Assoc. Petroleum Geologists Bull., **65**, 438–465.

Weber, K. J., 1971, Sedimentological aspects of the oil fields of the Niger Delta: Geol. en Mijnb., **50**, 559–576.

Weimer, R. J., and Land, C. J., Jr., 1972, Field guide to Dakota Group Cretaceous stratigraphy, Golden, Morrisson area, Colorado: Mountain Geologist, **9**, 241–268.

Weimer, R. J., Howard, J. D., and Lindsay, D. R., 1982, Tidal flats: in Scholle, P. A., and Spearing, D., eds., Sandstone depositional environments: Am. Assoc. Petroleum Geologists Mem. **31**, 191–246.

Wells, J. T., 1982, Season cycles of fine-grained sediment transport, west coast of Korea (Abs.): Geol. Soc. America Abstracts with Programs, **14**, 644.

———, 1983, Dynamics of coastal fluid muds in low-, moderate- and high-tide range environments: Can. Jour. Fisheries and Aquatic Sci., **40**, 130–142.

Wells, J. T., and Huh, O. K., 1980, Tidal flat muds in the Republic of Korea: Chinhae to Inchon: ONR Scientific Bull., **4**, 21–30.

Wells, J. T., Prior, D. B., and Coleman, J. M., 1980, Flowslides in muds on extremely low-angle tidal flats, northeastern South America: Geology, **8**, 272–275.

Wescott, W. A., and Ethridge, F. G., 1980, Sedimentology and tectonic setting of fan deltas with special reference to the Yallahs Fan Delta, southeast Jamaica: Am. Assoc. Petroleum Geologists Bull., **65**, 374–399.

——— and ———, 1983, Eocene fan delta-submarine fan deposition in the Wagwater Trough, east-central Jamaica: Sedimentology, **30**, 235–247.

Weser, O., 1975, Exploration for deep water sandstones: New Orleans Geol. Soc. Short Course Syllabus, 63 p.

Wessell, J. M., 1969, Sedimentary history of upper Triassic alluvial fan complexes in north-central Massachusetts: Univ. of Massachusetts Dept. of Geology Contrib. 2, 157 p.

Whitaker, J. H. M., 1974, Ancient submarine canyons and fan valleys: in Dott, R. H., Jr., ed., Modern and ancient geosynclinal sedimentation: Soc. Econ. Paleontologists and Mineralogists Spec. Pub. 19, 106–125.

White, S. M., Chamley, H., Curtis, D. M., Klein, G. deV., and Mizuno, A., 1980, Sediment synthesis: Deep Sea Drilling Project Leg 58, Philippine Sea: in Klein, G. deV., Kobayashi, K. et al., 1980: Initial Reports of the Deep Sea Drilling Project, v. 58: Washington, DC, U.S. Government Printing Office, 963–1013.

Wilde, P., Normark, W. R., and Chase, T. E., 1978, Channel sands and petroleum potential of Monterey Deep Sea Fan, California: Am. Assoc. Petroleum Geologists Bull., **62**, 967–983.

Wilkinson, B. H., 1975, Matagorda Island, Texas: the evolution of a Gulf Coast barrier complex: Geol. Soc. America Bull., **86**, 959–967.

Williams, D. F., and Rust, B. R., 1969, The sedimentology of a braided river: Jour. Sedimentary Petrology, **39**, 649–679.

Williams, G. E., 1971, Flood deposits of the sand-bed ephemeral streams of central Australia: Sedimentology, **17**, 1–40.

Williamson, C. R., and Hill, D. R., 1981, Submarine fan deposition of the upper Cretaceous Winters Sandstone, Union Island Field, Sacramento County, California: in Siemers, C. T., Tillman, R. W., and Williamson, C. R., eds., Deep water clastic sediments: a core workshop: Soc. Econ. Paleontologists and Mineralogists Core Workshop 2, 77–115.

Wilson, I. G., 1972, Aeolian bedforms—their development and origins: Sedimentology, **19**, 173–210.

Wright, L. D., and Coleman, J. M., 1974, Mississippi River mouth processes: effluent dynamics and morphologic development: Jour. Geology, **82**, 751–778.

Wright, M. E., and Walker, R. G., 1981, Cardium Formation (U. Cretaceous) at Seebe, Alberta—storm-transported sandstones and conglomerates in shallow marine depositional environments below fair-weather wave base: Can. Jour. Earth Sci., **8**, 795–809.

Index